P9-CEK-781

Global Warming®
FOR
DUMMIES®

by Elizabeth May
Zoë Caron

John Wiley & Sons Canada, Ltd.

Global Warming For Dummies®

Published by
John Wiley & Sons Canada, Ltd.
6045 Freemont Blvd.
Mississauga, ON L5R 4J3
www.wiley.com

For general information on John Wiley & Sons Canada, Ltd., including all books published by Wiley Publishing Inc., please call our warehouse, Tel. 1-800-567-4797. For reseller information, including discounts and premium sales, please call our sales department, Tel. 416-646-7992. For press review copies, author interviews, or other publicity information, please contact our publicity department, Tel. 416-646-4582, Fax 416-236-4448.

Library and Archives Canada Cataloguing in Publication Data

May, Elizabeth
Global warming for dummies / Elizabeth May, Zoë Caron.

Includes index.
ISBN 978-0-470-84098-6

1. Global warming—Popular works. I. Caron, Zoë II. Title.
QC981.8.G56M39 2008 363.738'74 C2008-902111-8

Printed in Canada on chlorine-free paper made from 100% post-consumer waste

1 2 3 4 5 FP 13 12 11 10 09

About the Authors

Elizabeth E. May has been recognized by the United Nations for her work in the environmental movement, both in June 1990 with the Global 500 Role of Honor for Environmental Achievement and on International Women's Day, March 8, 2006 by the United Nations Environment Program as one of the leading women in environment globally. Since 1997, she has served as a Commissioner in the Earth Charter Commission, co-chaired by Mikhail Gorbachev and Maurice Strong. Dr. May assisted in organizing the first international, comprehensive scientific conference into the climate change threat, in June 1988, hosted by Canada. She was engaged in the negotiation of the Montreal Protocol as Senior Policy Advisor to Canada's Minister of the Environment. She was a member of the International Policy Advisory Committee, World Women's Congress for a Healthy Planet, Miami, November 1991, served as an advisor in many capacities in the preparation for the United Nations Conference on Environment and Development (the Earth Summit), and was a board member for nine years for the International Institute for Sustainable Development.

Dr. May was Executive Director of Sierra Club of Canada for seventeen years, before leaving that position in 2006 to enter politics. She is currently the Leader of the Green Party of Canada. Dr. May is a lawyer, an author of six published books on Canadian environmental issues, and, most importantly, a mother and grandmother. Among many prestigious Canadian awards and honors, Dr. May has received the highest citizen honor in Canada, the Order of Canada, at the Officer level.

Zoë Caron has worked on initiatives to green university campuses through the Sierra Youth Coalition's Sustainable Campuses and the Energy Action Coalition's Campus Climate Challenge. Zoë is a founding member of the Canadian Youth Climate Coalition and past youth delegate to United Nations Climate Change Conferences in Canada and Kenya.

Identified as an emerging leader on climate change, she was profiled among colleagues as "The Next Generation" in Vanity Fair's 2007 Green Issue. She works with Students on Ice Expeditions, bringing students from around the world to the Arctic and Antarctic to learn about the importance of these regions to the rest of the planet. She currently writes for the Green Content Creation Group and serves on the Board of Directors of the Sierra Club of Canada.

Dedication

Elizabeth dedicates this book to her children and grandchildren, in hopes that by the time the youngest of you is old enough to read this book, the prognosis will be very different and far more hopeful.

Zoë dedicates this book to every individual who has dedicated her or him self to advocate climate change issues since the 1980s to bring the global community to the level of awareness we are at today.

And it goes without saying that they both dedicate this book to you, the reader, for making the choice to read about climate change.

Authors' Acknowledgments

Zoë and Elizabeth both want to express deep appreciation to many friends and colleagues who assisted in the research and writing of this book. A special thank you to Dr. Ian Burton, Dr. Jim Bruce, and Dr. Gordon McBean, leading scientists of the Intergovernmental Panel on Climate Change, who helped ensure the accuracy of the information presented in this book. Deep thanks to How-Sen Chong, founder of Carbonzero, for endless fact-checks and data provisions. We're grateful to those that have helped review chapter content: Dr. O.W. Archibold of the University of Saskatchewan; Dr. Jonathan Newman of the University of Guelph; Peter Howard of Zerofootprint; Ruth Edwards of the Canada Climate Action Network; Kristopher Stevens of the Ontario Sustainable Energy Association; and to the David Suzuki Foundation team of Nick Heap, Paul Lingl, and Dale Marshall. (As always, any errors and omissions are the authors' alone.) Thanks also to key image providers, Dr. Max Boykoff of the Oxford University Centre for the Environment, and John Streicker of the Northern Climate Exchange.

To Debra Eindigeur, Elizabeth's Executive Assistant, for assistance in managing the manuscript and ensuring drafts changed hands in timely fashion. To Cendrine Huemer and Jaymini Bihka for their research work. Ongoing gratitude to the countless colleagues called on for advice, feedback, or data.

Zoë and Elizabeth also want to express their deepest gratitude to the seemingly endless patience of our editor Robert Hickey for his always-excellent advice, text maneuvering, and overall guidance. A big thanks to those who worked behind the scenes: editor Colleen Totz-Diamond, our copy editor Laura Miller, project coordinator Lindsay Humphreys, and our brand reviewers Rev Mengle, Zoë Wykes, and Jennifer Bingham.

Elizabeth wants to say that (once again, as in previous books) nothing would be possible without the extraordinary grace, patience, and support of her daughter, Victoria Cate May Burton. No one has ever had a better daughter, and few have known a better person.

Zoë wants to thank her ever-patient friend Lilith Wyatt for postponing their South America excursion and for the many locutorio visits throughout the four-month trek. She thanks Jessica Budgell for her always-there encouragement, and Zoë apologizes to all those from whom she took a rain check so that she could spend time with her laptop instead.

This book was made possible by people who — intentionally or not — provided the most timely, impromptu, and gracious writing locales: the owners of Coburg Coffee in Halifax and of Planet Coffee and Bridgeheads in Ottawa, Liz McDowell, Louise Comeau, parents Michael Fischer and Julie Caron, Panny Taylor, Candace Batycki, Adriane Carr and Paul George, Anjali Helferty and Roxanne Charlebois, Kathryn Kinley, and Reina Lahtinen.

Last and foremost, Zoë thanks Elizabeth for endless mentorship, teaching, and patience. Few others would take time to edit while running a federal political party, recovering in-hospital from surgery, or making lobster salad for sixty.

Publisher's Acknowledgments

We're proud of this book; please send us your comments through our Dummies online registration form located at `www.dummies.com/register/`.

Some of the people who helped bring this book to market include the following:

Acquisitions, Editorial, and Media Development

Editor: Robert Hickey

Project Manager: Elizabeth McCurdy

Project Editor: Lindsay Humphreys

Copy Editor: Laura Miller

Technical Reviews: O.W. Archibold, Jonathan Newman, Peter Howard, Kristopher Stevens, Paul Lingl, Dale Marshall, Nick Heap, Ruth Edwards

Cover photo: Marc Romanell/The Image Bank/ Getty Images

Cartoons: Rich Tennant (`www.the5thwave.com`)

Composition Services

Vice-President Publishing Services: Karen Bryan

Project Coordinator: Lynsey Stanford

Layout and Graphics: Reuben W. Davis, Nikki Gately, Melissa K. Jester, Ronald Terry

Proofreaders: Laura L. Bowman, David Faust, Jessica Kramer

Indexer: Christine Spina Karpeles

Special Help: Zoë Wykes, Rev Mengle, Jennifer Bingham, Carrie Burchfield

---

John Wiley & Sons Canada, Ltd.

Bill Zerter, Chief Operating Officer

Jennifer Smith, Vice-President and Publisher, Professional and Trade Division

Publishing and Editorial for Consumer Dummies

Diane Graves Steele, Vice-President and Publisher, Consumer Dummies

Joyce Pepple, Acquisitions Director, Consumer Dummies

Kristin A. Cocks, Product Development Director, Consumer Dummies

Michael Spring, Vice-President and Publisher, Travel

Kelly Regan, Editorial Director, Travel

Publishing for Technology Dummies

Andy Cummings, Vice-President and Publisher, Dummies Technology/General User

Composition Services

Gerry Fahey, Vice-President of Production Services

Debbie Stailey, Director of Composition Services

Contents at a Glance

Table of Contents

Introduction

On Monday, the newspapers tell you the ice caps are melting, and people everywhere are about to be swept off in a giant flood. On Tuesday, you hear a radio interview with a scientist who says global warming is all a hoax. Wednesday finds you standing in the grocery line, listening to people muttering about how strangely warm the weather has been outside recently. By Thursday, you just don't know what to think anymore.

Think of today as Friday — the day all these stray pieces come together right here in your hands, thanks to *Global Warming For Dummies.*

Global warming is already changing the environment, the economy, and people's ways of living. The changes aren't over, either, and the more that changes around the world, the more you have to understand what global warming is. But you know what? It's really quite exciting. Although global warming is connected to scary scenarios featuring soaring temperatures and worsening hurricanes and monsoons, it's also a link to a better future. Global warming is opening doors for the development of new types of fuels, leading the shift to reliable energy sources, and creating a vision of a greener tomorrow. And the best part? You're right in the middle of it all, helping to make those changes.

About This Book

Global Warming For Dummies is your guide to climate change. We use *climate change* and *global warming* interchangeably in this book, though they are slightly different things, as we discuss in Chapter 1. This book gives you the basics so that you can understand the problem, relate it to your daily life, and be inspired to start working on solutions to this complex and important issue.

In this book, we explain the concepts behind global warming clearly and simply by using the latest, most credible science, mainly from the Fourth Assessment Report of the Intergovernmental Panel on Climate Change (IPCC).The IPCC is a team of more than 2,000 scientists who assess peer-reviewed climate change science and compile the assessments into a number of reports. These reports are mainly to inform the politicians and bureaucrats at the United Nations' decision-making table, but anyone looking for detailed scientific information on climate change can read them. The IPCC is the most credible source of climate change information in the world today. (We discuss the IPCC in greater detail in Chapter 11.)

Although this book covers what global warming is and its impact on the world, *Global Warming For Dummies* isn't just about the science. The book also looks at a wide range of solutions to tackle climate change. We explore everything from the big-picture solutions that governments can implement to a slew of practical, can-do-it-today solutions for you at work, at home, and on the road.

Foolish Assumptions

We wrote this book assuming that you know zero, nil, zilch about global warming. You don't have to look up the definitions of big, ridiculous words or drag out your high school science textbook to read this book.

We also assume, however, that you know global warming exists, that you recognize humans contribute to this problem, and that you want to understand why global warming is happening.

How This Book Is Organized

This book is divided into six parts, covering everything you need to know about the causes and effects of global warming — and the solutions.

Part I: Understanding Global Warming

This part sorts out what global warming actually is. If you want to understand the science behind why the world's climate is changing, check out these chapters. We take a look at the infamous greenhouse gases and explore how they're changing the way the climate works. We also consider some of the other factors that are shaping the planet's climate and explain why scientists are almost entirely certain that humanity's production of greenhouse gases is heating up the atmosphere.

Part II: Tracking Down the Causes

Part II explains where all the greenhouse gases we talk about in Part I are coming from. Two major offenders exist: fossil fuels (¾ of the problem) and deforestation (¼ of the problem). In this part, we investigate where fossil fuels come from and why they have such a huge influence on the atmosphere. We also look at where and why deforestation is happening, and why it's a major cause of climate change. Finally, we investigate how businesses and individuals have unwittingly set climate change into motion through emitting greenhouse gases.

Part III: Examining the Effects of Global Warming

In this part, we look at how global warming is creating changes around the world. We review what has already happened because of climate change and consider what the future might hold. From natural disasters such as floods and storms, to mass extinctions in the animal world, to the heavy toll global warming could take on humanity, the picture's pretty grim if civilization keeps doing what it's doing. Fortunately, people can change direction — so keep reading!

Part IV: Political Progress: Fighting Global Warming Nationally and Internationally

Not everyone loves politicians, but in this part, we look at how their work can go a long way to help fight global warming. First of all, we consider how governments at every level — from presidents to mayors — can help cut back on greenhouse gas emissions. Then, we look at how countries can work together to tackle this truly global challenge. The economic challenge of global warming is particularly daunting for developing nations. In the last chapter of this part, we investigate developing nations' situations and see what steps they're taking to be part of the solution to global warming.

Part V: Solving the Problem

Solving global warming is requiring a lot of changes on a lot of different fronts, which is why this part is the longest in the book. Some of these changes are quite modest; most are quite major. First, civilization needs to shake its addiction to fossil fuels and find new, cleaner renewable energy sources. Happily, those energy sources are out there (and in this part).

Businesses and industries play a major role, too; we look at how they can cut back on their greenhouse gas emissions and make money, to boot. Another part of solving the problem is raising awareness. Non-government organizations have a big role to play in getting people's attention. So, too, does the media. Some of Hollywood's biggest stars are getting involved, as well, lending their stellar wattage to the cause and adapting conscientiously eco-green lifestyles. And most importantly, it's down to you, the reader, and the countless ways that you can contribute to the fight against global warming.

Part VI: The Part of Tens

No *For Dummies* book is complete without the Part of Tens. Think of these lists as quick little hits of global warming wisdom. We cover ten things you can do right now to fight global warming, profile ten inspiring people who are leading the charge against climate change, debunk ten myths about global warming, and offer ten great online resources. Flip to these chapters whenever you need a fast bit of information or a quick jolt of inspiration.

Icons Used in This Book

Throughout this book, you see little icons sprinkled in the left margin. These handy symbols flag content that's of particular interest.

When you see this icon, it means disagreement exists over the topic being discussed.

This icon marks feel-good stories and major advances in the fight against climate change.

This icon marks a piece of information that's important to know in order to understand global warming and the issues that surround it.

This icon marks paragraphs in which we talk about serious issues that humanity needs to deal with as soon as possible.

Don't worry about reading paragraphs with this icon. This icon flags material that we think is interesting, but might be a little too detailed for your tastes.

Ready to make a difference? This icon points you to simple solutions that can help you reduce your greenhouse gas emissions or become a part of a bigger solution.

Where to Go from Here

If you're entirely new to the subject of global warming, you likely want to read this book the old-fashioned way, starting at the beginning and working through to the end. If you already know something about the subject or want to find out more about a specific topic, you can just open this book up at any chapter and start reading.

Part I
Understanding Global Warming

In This Part . . .

If you have questions about the science behind global warming, this part is the place to start. We introduce you to greenhouse gases, explain why they're vital for life on Earth, and provide you with a blueprint that explains just how they're heating up the atmosphere. We also investigate why scientists are certain that greenhouse gases are the cause of the global warming that we're experiencing today, and we consider some of the other factors that could be contributing to climate change.

Chapter 1

Global Warming Basics

In This Chapter

- Getting to know what global warming is all about
- Figuring out what started climate change in the first place
- Investigating the changes global warming might bring
- Examining the role governments can play in fighting global warming
- Finding solutions to the problem

The phrase "global warming" hasn't been around long, but climate change, as it's also known, is nothing new. In fact, it has been a constant throughout history. Earth's climate today is very different from what it was 2 million years ago, let alone 10,000 years ago. Since the beginnings of the most primitive life forms, this planet has seen many different climates, from the hot, dry Jurassic period of the dinosaurs to the bleak, frozen landscapes of the ice ages.

Today, however, the planet's experiencing something new: Its climate is warming up very quickly. Scientists are certain that this change has been caused by emissions produced by human activities. By examining previous changes in the Earth's climate, using computer models, and measuring current changes in atmospheric chemistry, they can estimate what global warming might mean for the planet, and their projections are scary.

Fortunately, Earth isn't locked into the worst-case-scenario fate yet. By banding together, people can put the brakes on global warming. This chapter explains the essentials of global warming and what everyone can do to achieve a greener future.

Global Warming 101

When "global warming" became a household phrase, *greenhouse gases,* which trap heat in the Earth's atmosphere, got a bad reputation. After all, those gases are to blame for heating up the planet. But, as we discuss in Chapter 2, greenhouse gases in reasonable quantities aren't villains, they're heroes. They capture the sun's warmth and keep it around so that life is possible on Earth. The problem starts when the atmosphere contains too great an amount of greenhouse gases. (In Chapter 3, we look at how scientists have determined the correlation between carbon dioxide in the atmosphere and temperature.)

Heating things up with greenhouse gases

Human activities — primarily, the burning of fossil fuels (which we look at in the section "The Roots of Global Warming," later in this chapter) — have resulted in growing concentrations of carbon dioxide and other greenhouse gases in the atmosphere. As we explain in Chapter 2, these increasing quantities of greenhouse gases are retaining more and more of the sun's heat. The heat trapped by the carbon dioxide blanket is raising temperatures all over the world — hence, *global warming.*

So far, Earth has seen a 1.4-degree Fahrenheit (0.8-degree Celsius) increase in global average temperature because of increased greenhouse gases in the atmosphere. Unfortunately, the amount of greenhouse gases that human activities produce grows daily. So, humanity's current behavior is driving temperatures up at an alarming rate. Temperatures in polar regions, such as the Arctic, are experiencing temperature rises that are twice the global average.

Investigating other causes of global warming

Global warming is a very complex issue that you can't totally understand without looking at the ifs, ands, or buts. Scientists are certain that the rapid changes to climate systems are due to the build-up of greenhouse gases, and they can't explain the current rates of global warming without factoring in the impact of human greenhouse gas emissions. Other elements play a role in shaping the planet's climate, however, including the following:

- **Cloud cover:** Clouds are connected to humidity, temperature, and rainfall. When temperatures change, so can a climate's clouds — and vice versa.
- **Long-term climate trends:** The Earth has a history of going in and out of ice ages and warm periods. Scientific records of carbon dioxide levels in the atmosphere go back 800,000 years, but people can only give educated guesses about the climate earlier than that.
- **Solar cycles:** The sun goes through a cycle that brings it closer to or farther away from the Earth. This cycle ultimately affects the temperature of this planet, and thus the climate.

We go over these other issues in greater detail in Chapter 3.

The Roots of Global Warming

Just what are humans doing to release all those greenhouse gases into the atmosphere? You can pin the blame on two main offenses: burning fossil fuels and deforestation.

Fueling global warming

When you burn *fossil fuels,* such as coal and oil (named fossil fuels because they're composed of ancient plant and animal material), they release vast amounts of greenhouse gases (largely, but not exclusively, carbon dioxide), which trap heat in the atmosphere. Fossil fuels are also a limited resource — meaning that humanity can't count on them over the long term because eventually they'll just run out.

The fossil fuel that produces the most greenhouse gas emissions is coal, and burning coal to produce electricity is the major source of coal-related greenhouse gases. The second-worst offender is using gasoline and diesel for transportation, followed by burning oil to generate heat and electricity. In fact, if people could replace the coal-fired power plants around the world and switch away from the internal combustion engine, humanity would have most of the problem licked. (Check out Chapter 4 for more fossil fuel info and Chapter 13 for the scoop on energy alternatives.)

Heating up over deforestation

Forests, conserved land, and natural habitats aren't important just for the sake of saving trees and animals. Forests and all greenery are important players in keeping the climate in check. Plants take in the carbon that's in the atmosphere and give back oxygen, and older trees hold on to that carbon, storing it for the duration of their lives. By taking in carbon dioxide, they're significantly reducing the greenhouse effect. (See Chapter 2 for more about how plants help the Earth keep atmospheric carbon at a reasonable level.)

Unfortunately, much of the world's forests have been cut down to make way for farmland, highways, and cities. Deforestation is responsible for about a quarter of greenhouse gas emissions. Rainforests are especially good at soaking up carbon dioxide because they breathe all year round. Temperate forests, on the other hand, don't absorb much carbon dioxide over the winter, practically going into hibernation. (Chapter 5 has more about deforestation.)

Examining the Effects of Global Warming Around the World

This book should really be called *Climate Change for Dummies.* Although global warming is the common term for the climate changes that the planet's experiencing (and scientists agree that average global temperature will increase with the build-up of greenhouse gases), the term doesn't tell the whole story. The Earth's average surface temperature is going up, but some areas of the planet may actually get colder or experience more extreme bouts of rain, snow, or ice build-up. Consequently, most scientists prefer the term climate change. In the following sections, we look at how different places around the world will experience climate change.

We want to warn you, much of this section is pretty depressing. But nothing is exaggerated — the information here is all based on peer-reviewed scientific reports. Just how serious could the global impact of climate change be? The first global comprehensive scientific conference, which was held in Toronto, Canada, in 1988, described the climate change issue in this way: "Humanity is conducting an unintended, uncontrolled, globally pervasive experiment whose ultimate consequences could be second only to a global nuclear war."

The United States and Canada

In the U.S. and Canada, average temperatures have been rising because of climate change. As a result, the growing season has lengthened; trees have been sucking in more carbon, and farms have been more productive. The warmer weather hasn't been all good news, however. Many plants and animals are spreading farther north to adapt to climate changes, affecting the existing species in the areas to which they're moving. And increased temperatures have already been a factor in more forest fires and damage by forest insects, such as the recent pine beetle epidemic in the interior of British Columbia, Canada. (See Chapter 8 for more information about how global warming will affect animals and forests.)

Scientists project that the U.S. and Canada will feel the effects of climate change more adversely in the coming years. Here are some of the problems, anticipated to only get worse if civilization doesn't dramatically reduce greenhouse gas emissions:

- **Droughts:** Rising temperatures are increasing droughts in areas that are already arid, putting even larger pressure on scarce water sources in areas such as the U.S. Southwest.

- **Evaporating lakes:** The cities in the great heartland of the Great Lakes Basin will face retreating shorelines when the water levels of the Great Lakes drop because of increased evaporation. Lower water levels will also affect ship and barge traffic along the Mississippi, St. Lawrence, and other major rivers.
- **Floods:** Warmer air contains more moisture, and North Americans are already experiencing more sudden deluge events, causing washed out roads and bridges, and flooded basements.
- **Major storms:** Warming oceans increase the risk of extreme weather that will plague coastal cities. Think of Hurricane Katrina, arguably the most devastating weather event ever to hit a North American city. In a possible sign of what's to come, the super-heated waters of the Gulf of Mexico whipped Katrina into a hurricane with a massive punch while it crossed from Florida and made landfall in Louisiana, ultimately doing more than $200 billion worth of damage.

  REMEMBER

  Not all extreme weather events are hurricanes. Global warming is expected to increase ice storms in some areas and thunderstorms in others.
- **Melting glaciers:** Glaciers in the Rockies and in the far North, in both Canada and the U.S., are in retreat. Glacier National Park could some day be a park where the only glacier is in the name. When glaciers go, so does the spring recharge that flows down into the valleys, increasing the pressure on the remaining water supplies. People who depend on drinking water from rivers or lakes that are fed by mountain glaciers will also be vulnerable.
- **Rising sea levels:** Water expands when it gets warmer, so while global average temperatures rise, warmer air warms the ocean. Oceans are expanding, and sea levels are rising around the world, threatening coastal cities — many of which are in the U.S. and Canada. This sea level rise could become far more devastating should ice sheets in Greenland and Antarctica collapse.

Changes across northern Canada and Alaska are more profound than in the south. We discuss these impacts in the section "Polar regions," later in this chapter.

On average, North Americans have many resources, in comparison to developing regions of the world, to help them adapt to climate change. The Intergovernmental Panel on Climate Change (IPCC) says Canada and the United States can take steps to avoid many of the costs of climate change, to better absorb the effects and avoid the loss of human lives. For example, North America could establish better storm warning systems and community support to make sure that poor people in inner cities have some hope of relief during more frequent killer heat waves. (See Chapter 10 for more information about what governments can do to help their countries adapt to the effects of climate change.)

Latin America

South America has seen some strange weather in the past few years. Drought hit the Amazon in 2005, Bolivia had hail storms in 2002, and the torrential rainfalls lashed Venezuela in 1999 and 2005. In 2003, for the first time ever, a hurricane hit Brazil. Were these strange weather events linked to global warming? Scientists can't say with certainty, but these events are the kind expected to occur because of climate change. Scientists do project that extreme weather caused by climate change will increase. Events such as these may be signs of what's to come for Latin America.

Other changes in Latin America may be attributable to global warming. Rain patterns have been changing significantly. More rain is falling in some places, such as Brazil, and less in others, such as southern Peru. Glaciers across the continent are melting. This glacier loss is a particular problem in Bolivia, Ecuador, and Peru, where many people depend on glacier-fed streams and rivers for drinking water and electricity from small-scale hydroelectric plants. (See Chapter 9 for more about how global warming will affect humans.)

Scientists project that the worst is yet to come. The IPCC models anticipate that about half of the farmland in South America could become more desert-like or suffer saltwater intrusions. If sea levels continue to rise at a rate of 0.08 to 0.12 inches (2 to 3 millimeters) per year, it could affect drinking water on the west coast of Costa Rica, shoreline tourism in Mexico, and mangrove swamps in Brazil.

The threat to the Amazon from logging and burning has attracted the concern of celebrities such as Sting and Leonardo DiCaprio. But human-caused global warming could potentially do more damage than loggers. By mid-century, the IPCC predicts that parts of the Amazon could change from wet forest to dry grassland. Such a radical alteration would have an enormous impact on the millions of species of plants and animals that make this rich ecosystem their home. (We cover how ecosystems will be affected by climate change in Chapter 8.)

Europe

Recent findings have shown that climate change is already well under way in Europe. Years ago, the IPCC projected the changes that the continent is experiencing today: rising temperatures, increased intensity and frequency of heat waves, and increased glacier melt.

As for what's in store for Europe, the IPCC reports a 99-percent chance that Europe will experience other unfavorable climate changes. These changes may include the following:

- Increased occurrence of rock falls in some mountainous regions because of melting permafrost loosening mountain walls.
- More flash floods in inland areas.
- More heat waves and droughts in central, eastern, and southern Europe. This will have an impact on health and tourism in southern Europe in particular.
- Rising sea levels, which will increase erosion. These rising sea levels, coupled with storm surges, will also cause coastal flooding. The Netherlands and Venice will have greater trouble than other areas in Europe dealing with the rising sea level.
- The loss of up to 60-percent of Europe's native species of plants and animals. Fisheries will also be stressed.

GOOD NEWS

Countries sitting on the North Atlantic are likely to see a growth in fisheries, according to the IPCC.

These possibilities are all serious, but none of them represents the worst-case scenario — the Gulf Stream stalling. The results of this (stopping of a major ocean current) would be disastrous for Europe. (We look at the Gulf Stream in Chapter 7.) Although the Gulf Stream stalling is possible, the IPCC doesn't consider this possibility likely.

Africa

On a per-person basis, Africans have contributed the very least to global warming because of overall low levels of industrial development. Just look at a composite photo of the planet at night: The U.S., southern Canada, and Europe are lit up like Christmas trees, burning energy that results in greenhouse gas emissions. Africa, on the other hand, shows very few lights: some offshore oil rigs twinkle, and a few cities shine, but the continent is mostly dark.

Despite contributing very little to the source of the problem, many countries in Africa are already experiencing effects of global warming. Long periods of drought followed by deluge rainfall have had devastating impacts in places such as Mozambique. Coastal areas in East Africa have suffered damage from storm surges and rising sea levels.

Unfortunately, because of pervasive poverty and the scourge of AIDS, many areas of Africa lack the necessary resources to help people living there cope with climate change. And the effects of global warming may act as a barrier to development and aggravate existing problems. At present, 200 million people (or 25 percent of the continent's population) lack drinkable water. Climate change may boost this figure by another 75 to 200 million people over the next 12 years. The IPCC projects that some countries could see a 50-percent drop in crop yields over the same period and a 90-percent drop in revenue from farming by the year 2100. (We look at how developing nations are affected by and are addressing global warming in Chapter 12.)

Asia

More people call Asia home than any other continent — 4 billion in all. This high population, combined with the fact that most of Asia's countries are developing, means that a lot of people won't be able to sufficiently adapt to climate change impacts. As in Africa, climate change may bring pressures to the continent that will slow down development.

The first concern is the future availability of drinkable water, which is already under pressure from population growth, pollution, and low or no sanitation. The IPCC projects that anywhere from 120 million to 1.2 billion people may find themselves without enough drinkable water within the next 42 years, depending on the severity of climate change. Already, rising temperatures are causing glaciers in the Himalayas, which supply water to 2 billion people, to melt. These disappearing glaciers are aggravating water shortages and will increase avalanches and flooding.

Rising sea levels will be a concern for coastal Asia. The IPCC reports that mangroves, coral reefs, and wetlands will be harmed by higher sea levels and warming water temperatures. On a brighter note, with rising saltwater intruding into fresh water, both fish that thrive in slightly salty water and the industry that fishes them are expected to benefit. Unfortunately, this slightly salty water won't be good for freshwater organisms, as a whole. (See Chapter 8 for more about the impact global warming will have on the oceans.)

Illnesses in Asia are also expected to rise because of global warming. Warmer seawater temperatures could also mean more, and more intense, cases of cholera. Scientists project that people in South and Southeast Asia will experience more cases of diarrheal disease, which can be fatal. (Chapter 9 offers more information about how global warming might increase disease.)

Australia & New Zealand

If you ask an Australian about global warming, you probably won't get any argument about its negative effects. Australia has already experienced increased heat waves, less snow, changes in rainfall, and more than seven years of persistent drought in four of its states. This heat and lack of precipitation will likely worsen while global warming's effects intensify. According to the IPCC, by 2030, water will be even scarcer in southern and eastern Australia and northern New Zealand than it currently is.

Climate change has also strongly affected the ocean. Sea levels have already risen 2.8 inches (70 millimeters) in Australia since the 1950s, and increasing ocean temperatures threaten the Great Barrier Reef. The reef is at risk of bleaching, and the possibility that it may be lost altogether is becoming more real. (See Chapter 8 for details.)

As in the United States and Canada, however, Australia and New Zealand may see some benefits from climate change. The IPCC reports that the countries may experience longer growing seasons, fewer occurrences of frost, and increased rainfall in some regions, which would benefit farming and forestry.

Small islands

You probably aren't surprised to hear that when it comes to climate change, rising sea levels and more extreme storms create an enormous risk for small islands everywhere, such as the South Pacific island of Tuvalu. Some may simply disappear due to rising sea levels. If sea level rise does not inundate the islands, the impacts could still be severe:

- **Forests vulnerable to major storms:** Storms can easily topple island forests because a forest's small area doesn't provide much of a buffer (although some forests will expand because of warmer temperatures).
- **Limited resources:** Some islands can't adapt physically and/or financially.
- **Proximity of population to the ocean:** At least 50 percent of island populations live within a mile (1.5 kilometers) of water, and these populations are threatened by rising sea levels.
- **Risks to drinkable water:** The intrusion of ocean saltwater because of rising sea levels could contaminate islands' drinkable water, which is already limited on most islands.
- **Reliance on tourism:** Beach erosion and coral reef damage, two possible effects of climate change, would undermine tourism, which many islands rely on for their source of income.
- **Vulnerable agriculture:** Island agriculture, often a key part of the local economy, is extremely susceptible to harmful saltwater intrusions, as well as floods and droughts.

Polar regions

The planet's polar regions are feeling climate change's effects more intensely than anywhere else in the world. Warming temperatures are melting the ice and the permafrost that used to be solid ground.

The Arctic is home to many changes brought on by global warming, including the following:

- **Lost traditions:** Some indigenous people who make their homes in the Arctic are having to abandon their traditional ways of life. The Arctic ice and ecosystem are both core to many of these people's cultures and livelihoods. For more on this issue, flip ahead to Chapter 9.
- **Melting ice:** The Greenland ice sheet is quickly melting, adding to sea level rise. Arctic ice is also steadily losing ice volume. All of this melting is diluting ocean waters, affecting ocean currents.
- **New plant life:** Greenery and new plants have been appearing in the Arctic in recent years. The tree line, which used to end about ¾ of the way up Canada and Russia, is shifting farther north, but the soil is not there to support a forest.

Some people look forward to the changes that the Arctic is experiencing. Now that so much sea ice has melted, ships can navigate the Arctic Ocean more efficiently, taking shorter routes. Industrialists keenly anticipate being able to reach more fossil fuels below what used to be unreachable riches because of ice cover. Communities in the Arctic may be able to harness river flows that have been boosted or created by ice melt to run hydroelectric power. These short-term economic developments cannot outweigh the negative planetary impacts.

In the Antarctic, some scientists project major change because of global warming, thinking there's a chance that the western Antarctic ice sheet might melt by the end of the 21st century. The western Antarctic ice sheet is simply enormous. It contains about 768,000 cubic miles (3.2 million cubic kilometers) of ice, about 10 percent of the world's total ice. It appears to be weakening because warmer water is eroding its base. Most scientists dispute the notion that the entire sheet will melt, and many scientists are still researching the situation. Nevertheless, parts of the western Antarctic ice sheet are definitely melting, even if the whole thing isn't yet. The Greenland ice sheet is also melting — quickly. Both the western Antarctic and Greenland ice sheets are adding to sea level rise.

The melting polar ice is also endangering many species, such as polar bears and penguins, which rely on the ice as a hunting ground. (Chapter 8 offers more information about the ways the polar animals are being affected by global warming.)

Positive Politics: Governments and Global Warming

Governments are often the first institutions that the public looks to for big solutions. Governments represent the people of a region, after all, and are expected to make decisions for the good of the public. So, governments need to be able to respond to global warming effectively. Climate change is a very big problem for which no one has all the answers. Despite this challenge, governments around the world are willing to play their part — and it's an important one.

Making a difference from city hall to the nation's capital

All levels of government, from cities and towns, to states and provinces, to countries, have the ability to affect taxes and laws that can help in the fight against climate change:

- **Local governments:** Can implement and enforce city building codes, improve public transit systems, and implement full garbage, recycling, and composting programs.
- **Regional governments:** Can set fuel efficiency standards, establish taxes on carbon dioxide emissions, and set efficient building codes.
- **Federal governments:** Can lead on the largest of issues, such as subsidizing renewable energy sources, removing subsidies from fossil-fuel energy sources, taxing carbon, and developing national programs for individuals who want to build low-emission housing. Federal government can also set standards and mandatory targets for greenhouse gas reductions for industry, provinces, and states to follow.

The most effective governments work with each other — partnerships between cities, states, and countries exist around the world, supporting one another while they work on the same projects. To read more about what governments can do and are already doing, check out Chapter 10.

Working with a global government

Countries must work together through global agreements to deal with, and conquer, a problem as urgent, complex, and wide-sweeping as climate change. Global agreements create a common level of understanding and allow countries to create collaborative goals, share resources, and work with each other towards global warming solutions. No one country can solve climate change on its own, just like no one country created global warming in the first place.

The core international law around climate change is the UN Framework Convention on Climate Change (UNFCCC) and its more detailed agreement, the Kyoto Protocol. Countries have agreed that by the end of 2012, they'll have collectively reduced global greenhouse gas emissions by about a third of what they are today. Some countries have done their part, others have done little, and some have exceeded expectations. The international discussions are ongoing; government representatives meet on an annual basis for the United Nations Climate Change Conference. We discuss just what goes on at those meetings in Chapter 11.

Helping developing countries

The effects of climate change are taking a particularly heavy toll on the populations of *developing countries* — countries with little or no industry development and a weak or unstable economy. These countries, which are primarily located in Latin America, Asia, and Africa, have fewer financial resources to recover from events such as flooding, major storm damage, and crop failures. Money that these nations have to spend paying for the effects of global warming is money that they can't spend building their economies.

Developing countries have little or no major industry development, for the most part (although China is now just overtaking the U.S. as the world's largest polluter), so they don't add many greenhouse gas emissions to the atmosphere. Even China, with its growing industry, lags far behind the emissions of industrialized nations on a per-person basis. Because industrialized countries have been the primary greenhouse gas emitters, global consensus is that they're firstly responsible for reducing emissions, and they can also play a role in helping developing countries shift to renewable energy sources and adapt to climate impacts. For more about how developing nations are addressing climate change, see Chapter 11.

Solving the Problem

Everyone can play a part in slowing down global warming, and humanity doesn't have time to start small. Solving climate change requires a major commitment from everyone — from big business and industry to everyday people. Combined, these changes can make the necessary difference.

Changing to alternative energies

Fossil fuels (see Chapter 4) are the primary source of the human-produced greenhouse gases causing global warming. Although they've fuelled more than a century of human progress, it's time to leave them with the dinosaurs. Fortunately, a wide array of energies is waiting to take the place of oil, coal, and gas. Some of these energy sources aren't yet ready for modern civilization to use them on a grand scale, but if businesses and governments commit to developing these energies, they soon will be.

Here's a list of *renewable resources* — energy that doesn't run out, unlike fossil fuels, and doesn't pump more carbon into the atmosphere:

- **Geothermal:** Jules Verne was wrong; the center of the planet doesn't contain another world, but it does have plenty of heat. People can use that heat to boil water to produce steam that propels turbines and generates electricity. Even areas without geo-heat sources to boil water can heat homes through *geothermal* energy (the warmth of the earth).
- **Hydro:** People can harness *hydropower,* or water power, to turn turbines and create electricity.
- **Solar:** Humanity can use the sun's warmth in a few ways. Solar cells, like you see on some roofs, can convert sunlight to electricity. People can also heat buildings and water with the sun's direct heat.
- **Waste:** Garbage is more than just trash. It offers astounding possibilities. People can harness the methane emitted from dumps, burn the byproducts of agriculture as fuel, and even use old frying oil as a type of diesel.
- **Wind:** Remember that pinwheel you had as a kid? Giant versions of those wheels are popping up all over the world as wind turbines, generating clean electricity for homes, businesses, and entire energy grids.

Feeling charged up? Check out Chapter 13 to further explore the renewable-energy possibilities.

Getting down to business

Industries are the largest contributors to greenhouse gas emissions, and they can make the biggest contributions to the fight against global warming. Although some of the changes that businesses can make may have an initial impact on the businesses' pocketbooks, many of those changes might even save businesses money in the long run.

Industrial-strength solutions

The greatest immediate change businesses and industries can make is to improve their efficiency. Companies waste a lot of energy powering antiquated equipment, heating poorly insulated buildings, and throwing out materials that they could recycle. Chapter 14 details some of the ways that companies can pull up their socks and make smarter use of energy, and it also shares some impressive success stories.

Ideally, renewable energy will ultimately power industry. Unfortunately, much of the energy that industry uses still comes from fossil fuels for the time being; not all the alternative energies that we describe in the section "Changing to alternative energies," earlier in this chapter, are mature enough to power major factories. However, industry doesn't need to pump the greenhouse gas emissions from those fossil fuels into the air. Currently, some scientists and industries are storing carbon emissions underground. This solution is controversial, however. We consider the issue in Chapter 13.

Green fixes for forestry and farming

The forestry and agriculture industries can do more than just cut back on their greenhouse gas emissions; they can actually increase the amount of carbon that's absorbed from the planet's atmosphere. (See Chapter 2 to take a ride on the carbon cycle and understand the critical role that plants play in keeping Earth livable.)

Around the world, forests are being cut down, removing valuable *carbon sinks,* which absorb carbon from the atmosphere. Where they harvest trees, logging companies need to explore methods other than clear-cutting; selectively harvesting trees enables forests to continue to thrive. In other countries, particularly in South America, people are clearing forests for farmland. Losing those forests is particularly costly for the atmosphere because, unlike forests in more temperate climates, these rainforests absorb carbon year-round. Deforestation methods have to change.

Farming's solution for global warming is dirty — or how dirt is treated. Believe it or not, a simple action like excessively tilling the land causes carbon to be released into the atmosphere. And when farmers add greenhouse gas–laden fertilizers to the soil, they release even more emissions. By cutting back on tilling the land and using less fertilizer, farmers can be a potent part of the solution to climate change.

Making it personal

You're a vital part of the climate change solution, too. As a citizen, you can ensure that governments recognize the importance of global warming and follow through on their promises. As a consumer, you can support companies that are making the biggest strides in fighting climate change and encourage other companies to make reducing greenhouse gases a priority. If you're really passionate about having your voice heard, you might even want to consider joining a group dedicated to spreading the word about global warming. We tell you how you can get involved in Chapter 15.

You can also make many changes in your daily life — some that seem small, some less so — that cut back on the carbon emissions for which you're responsible. You're probably already familiar with many of the little steps you can take to be more climate friendly:

- **Making your home more energy efficient:** Better insulate your roof, basement, and walls; seal your windows; and replace your old light bulbs with compact fluorescent bulbs.
- **Reducing the amount of garbage you produce:** Take a reusable bag with you when you shop, buy unpackaged goods, and recycle and reuse materials.
- **Using energy wisely:** Turn off lights and appliances when you're not using them, use the air conditioner less in the summer, and turn down the heat in the winter.

Did you know that many of your appliances are gobbling electricity, causing the emission of greenhouse gases, even when those appliances are turned off? Or that putting a lid on pots on your stove makes your food cook more efficiently? Chapter 18 offers all sorts of tidbits on how you can go green easily.

Not every action that you can take to cut back on your greenhouse gas emissions is manageable — not everyone can buy a hybrid car (which we consider in Chapter 17) or build a home that doesn't rely on major power producers for energy (see Chapter 18) yet. But, hopefully, we suggest some options in this book that fit your situation and can help you to make a difference.

Global warming affects everyone, and everyone can play an important role in stopping it. Forget the doom and gloom you may hear on the news — start thinking about the exciting opportunities you have to make a change.

Chapter 2

The Greenhouse We Live In

In This Chapter

- Understanding the greenhouse all humanity lives in
- Taking a close look at carbon dioxide
- Checking out the other greenhouse gases

When you watch the news or read the papers, you probably hear a lot about the greenhouse effect and greenhouse gases. What you've read or seen might make you think that the greenhouse effect and greenhouse gases are all bad.

In fact, greenhouse gases have long been the good guys. They're necessary for life on Earth. When it comes to humanity's survival (and the survival of everything else on Earth), humans need their home planet to be like the porridge in *Goldilocks and the Three Bears:* not too hot or too cold, but just right. A planet with too few greenhouse gases would be too cold; an atmosphere with a lot of greenhouse gases traps too much heat.

Greenhouse gases become a problem only when the atmosphere contains too much of them, which is happening today. Our industries and our farms, even our garbage dumps, are pumping out an array of gases — carbon dioxide, methane, nitrous oxide, and a host of other substances. Humanity has knocked off kilter the life-preserving cycle that makes sure the Earth's atmosphere has just enough carbon dioxide, the star greenhouse gas.

The Greenhouse Effect 101

If you want to understand the greenhouse effect, the best place to start is with the object that gave us this analogy in the first place, the greenhouse. A greenhouse works by letting in sunlight, which plants and soil absorb, thus heating up the greenhouse. The panes of glass ensure that the warmer air doesn't escape the greenhouse, or does so very slowly. If you've ever parked your car with the windows rolled up on a sunny day, you've experienced this effect. When you open your car door, you're hit with a blast of hot air. The windows of your car have acted like the panes in a greenhouse, letting sunlight in, which heats the car, and then trapping that heat.

Certain gases in our atmosphere trap the sun's heat in a similar way. These particular gases are called *greenhouse gases* because they cause this greenhouse effect. The Earth is bombarded by radiation from the sun. Some of this radiation can be seen (think visible light), and some of it can't be (ultraviolet light, for example).

Very hot bodies give off different amounts of energy than cold ones do. A basic law of physics says that everything gives off *radiative* (mostly heat or light) energy, and how much energy it emits depends on its temperature. The sun, for example, is a toasty 10,300 degrees Fahrenheit (5,700 degrees Celsius) — a little bit hotter than people are used to here on Earth. So, the sun gives off a lot of radiative energy, and the Earth gives off very little. Earth is warm mostly because of the heat it gets from the sun — most of the sun's radiative energy actually zooms right through the atmosphere to the Earth's surface. (The helpful high level ozone layer protects us by absorbing a lot of the harmful ultraviolet rays.)

A portion of this radiation, about 30 percent on average, bounces off clouds, ice, snow, deserts, and other bright surfaces, which reflect the sun's rays back into outer space. The other 70 percent is absorbed by land or water, which then heats up. And the Earth emits some of that heat — in the form of *infrared radiation* (electromagnetic waves most commonly known as heat). The unique qualities of the greenhouse gases come into play: The greenhouse gases absorb some of the escaping infrared radiation in the lower atmosphere, warming the Earth, thus creating the greenhouse effect.

So, less of the radiation from the Earth's surface gets to outer space than it would have without those gases, and that energy remains in the atmosphere and returns to the Earth's surface — making both the atmosphere and Earth itself warmer than they would be otherwise.

To see the greenhouse effect in action, take a look at Figure 2-1.

If the planet had no atmosphere or greenhouse gases, humanity would be left out in the cold. The Earth wouldn't be able to keep any of the heat that it gets from sun. Thanks to greenhouse gases, humanity is kept reasonably warm, enjoying an average temperature of 59 degrees Fahrenheit (15 degrees Celsius), some 62.6 to 64.4 degrees Fahrenheit (17 to 18 degrees Celsius) warmer than without greenhouse gases.

This natural greenhouse effect and the ozone layer allow life to exist on Earth. Without the greenhouse effect, the Earth would be too cold. And without the ozone layer, life couldn't survive the sun's ultraviolet radiation.

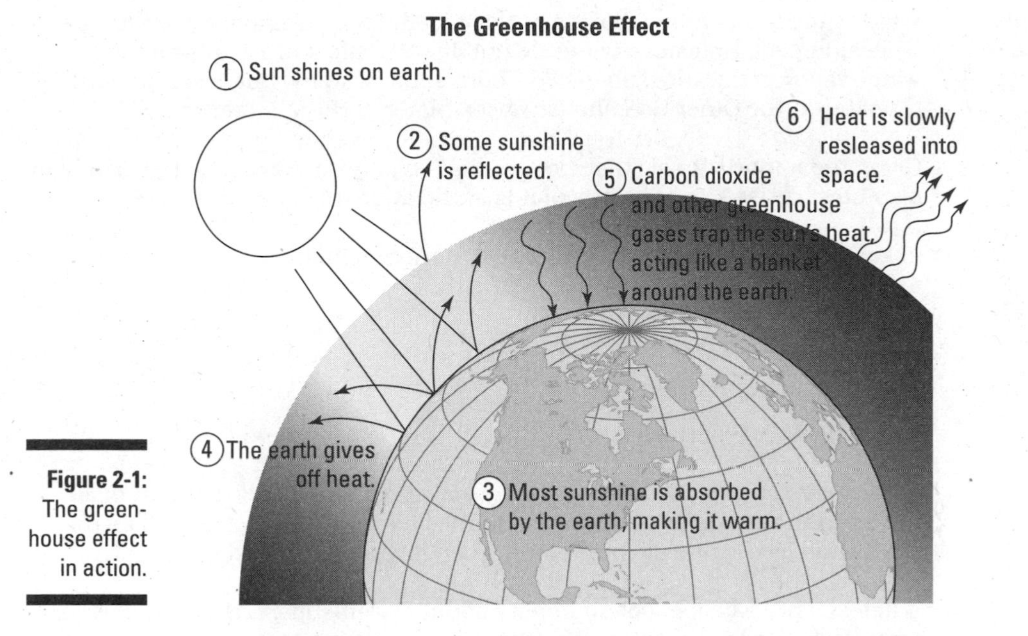

Figure 2-1: The greenhouse effect in action.

Too much greenhouse gas turns the heat up beyond that to which societies and ecosystems have become adjusted. The atmosphere on the planet Venus is 96 percent carbon dioxide (the key greenhouse gas that we talk about in the following section). Because of Venus's excessive amount of greenhouse gases, that cloudy molten planet experiences surface temperatures of up to 500 degrees Celsius. The atmosphere on Mars has 95 percent carbon dioxide but is very thin, and the planet's position is farther away from the sun than Earth — leaving it at a brisk –80 degrees Fahrenheit (–60 degrees Celsius).

Carbon Dioxide: Leader of the Pack

Earth's atmosphere contains 24 different greenhouse gases, but just one of them accounts for the overwhelming majority of the effect: carbon dioxide (or CO_2, for short). This gas accounts for about 63 percent of the greenhouse gas warming effect in the long run. (In the short term, over the last 5 years, it has accounted for 91 percent. See Table 2-1 for more on the intensity of gases over time.) If you're itching to know about the other 37 percent of greenhouse warming, check out the section "Looking at the Other Greenhouse Gases," later in this chapter.

Water vapor, not carbon dioxide, is technically the greenhouse gas with the biggest impact. Human activities do not directly affect in a significant way water vapor in the atmosphere. For more about water vapor, see the section, "Looking at the Other Greenhouse Gases," later in this chapter.

Given the important role that carbon dioxide plays in warming the Earth, you may be surprised by how little of it is in the atmosphere.

In fact, 99.95 percent of the air we breathe (not including water vapor) is made up of

- **Nitrogen:** 78 percent
- **Oxygen:** 21 percent
- **Argon:** 0.95 percent

Carbon dioxide, by contrast, currently makes up only 0.0385 percent of all the air in the atmosphere. Human activities have helped increase that concentration from pre-industrial times, when it was about 0.0280 percent.

When scientists talk about air quality and the atmosphere, they often use the term parts per million (ppm). So, out of every million parts of air, only 385 are carbon dioxide. That's not much carbon dioxide, but what a difference it makes! It's like the hot pepper you put into a huge pot of chili — this stuff is literally the spice of life, and just a pinch will do ya! And, like the pepper in the chili, if you have too much, watch out.

Looking at the carbon cycle

Carbon dioxide occurs naturally — in fact, you produce some every time you exhale. You inhale oxygen (and other gases), which your body uses as a nutrient, and you breathe out what your body doesn't need, including carbon dioxide. You aren't alone in using this process. Every animal and insect on Earth that breathes in air exhales carbon dioxide. But a lot of organisms, most of them plants, suck carbon dioxide out of the air. Trees, for example, take in carbon dioxide and give out oxygen — the complete opposite of what people do.

The *carbon cycle* is the natural system that, ideally, creates a balance between carbon emitters (such as humans) and carbon absorbers (such as trees), so the atmosphere doesn't contain an increasing amount of carbon dioxide. It's a huge process that involves oceans, land, and air. Life as we know it — from microscopic bugs in the oceans to you and me, and every fern in between — would disappear without it. You can think of the carbon cycle almost as the Earth breathing in and out.

A recipe that gives you gas

Carbon dioxide is composed of one carbon atom and two oxygen atoms, connected by double bonds. Present in our atmosphere as an odorless, colorless gas, it can also exist in solid form (think dry ice) and, when kept under pressure, in liquid form (the bubbles you see in champagne are carbon dioxide escaping after you uncork the bottle and remove the pressure).

The carbon cycle is called "in balance" when roughly the same amount of carbon that's being pumped into the air is being sucked out by something else. The atmospheric concentration of carbon dioxide was at a concentration of 280 ppm — carbon dioxide concentrations have fluctuated up and down, but 280 ppm has been about the highest recorded concentration for the past 800,000 years — until humans started to distort it. (We look at how humans contribute carbon dioxide to the atmosphere in Part II.)

When trees take up carbon through photosynthesis, they're called *carbon sinks.* Plants aren't the only carbon sinks, however. Figure 2-2 shows how the ocean, plants, and soil all act as carbon sinks, removing carbon from the atmosphere. They also store, or *sequester,* carbon, and they store the carbon in a carbon *reservoir.* For example, the ocean holds about 38,000 billion metric tons of carbon in its reservoir.

Under the deep blue sea

The ocean is the biggest carbon sink on Earth. So far, it has tucked away about 86 percent of all the carbon dioxide in the world. If that gas was in the atmosphere, not underwater, the world would be a lot hotter.

The exchange of carbon dioxide between the ocean and the air happens at the surface of the water. When air mixes with the surface of the ocean, the ocean absorbs carbon dioxide because carbon dioxide is *soluble* in water (that is, carbon dioxide can be absorbed by water). And, in fact, the seas' ability to absorb carbon dioxide is referred to as the *solubility pump* because it functions like a pump, drawing carbon dioxide out of the air and storing it in the ocean.

The ocean also acts as a *biological pump* to remove carbon dioxide from the atmosphere. Plants close to the surface of the ocean take in carbon dioxide from the air and give off oxygen, just like plants on land. (We discuss this process, known as photosynthesis, and the role that plants play in the carbon cycle in the following section.) *Phytoplankton* are microscopic plants that live in water. You may know them as algae, most commonly seen as the greenish clumpy plants that float around on ponds and other water. Phytoplankton have short but useful lives. If other organisms don't eat them, they simply die within just a few days. They then sink to the ocean floor, mix into the sediment, and decay. The carbon dioxide that these plants absorb during their brief lives is well and truly sequestered after their little plant bodies are buried.

Each year, the oceans put away about another 2 billion metric tons of carbon dioxide. Figure 2-3 demonstrates how the ocean interacts within the carbon cycle.

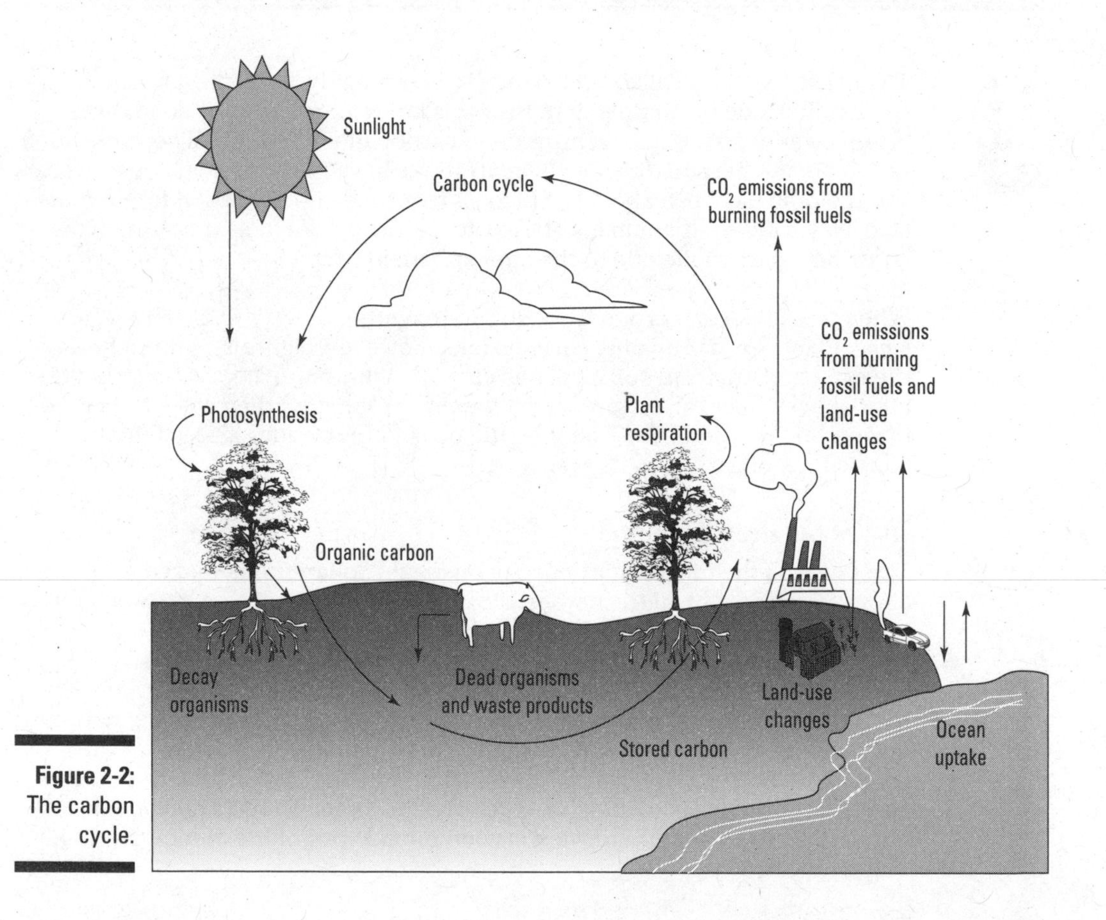

Figure 2-2: The carbon cycle.

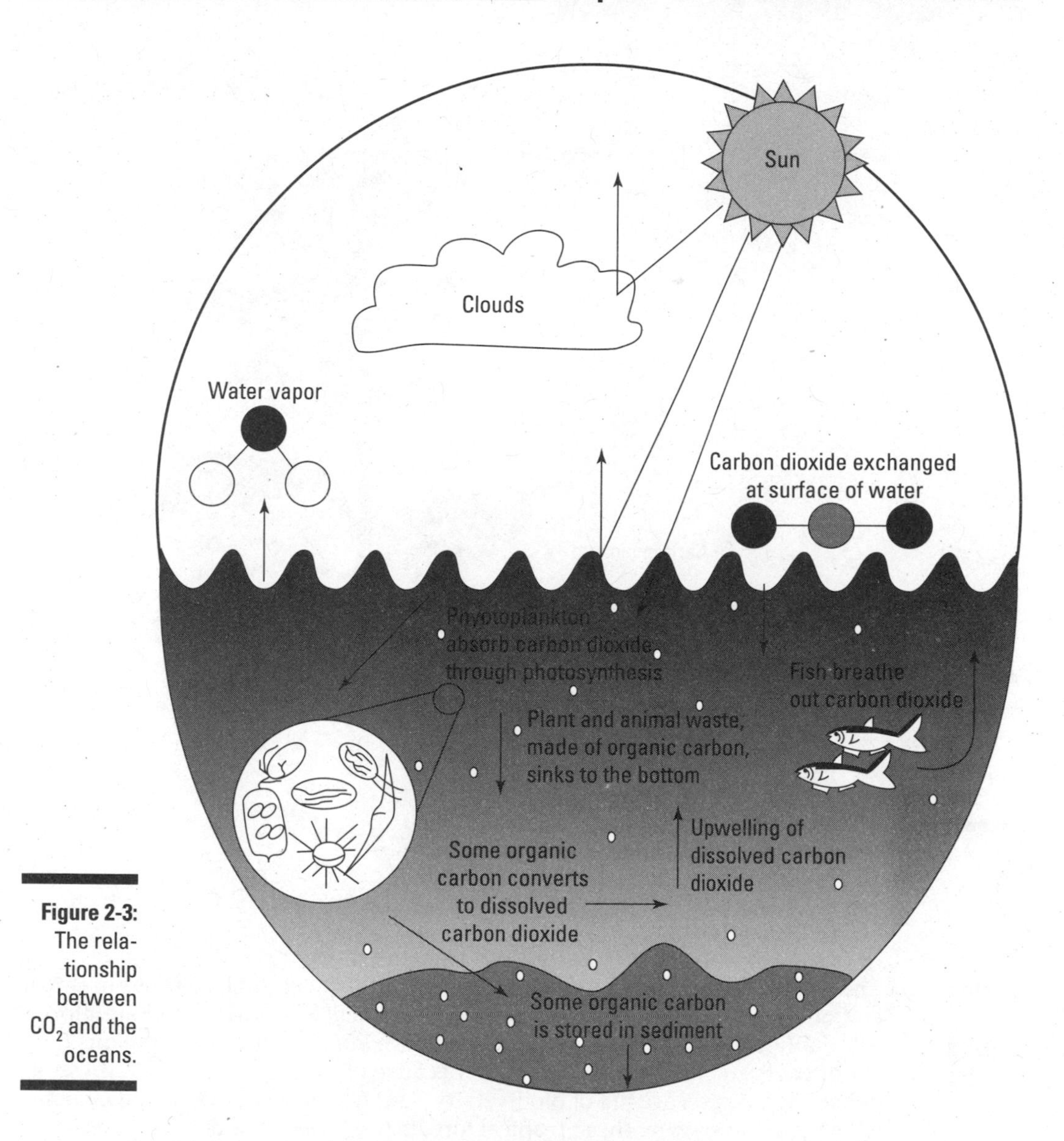

Figure 2-3: The relationship between CO_2 and the oceans.

Why people couldn't survive without plants

You may not have realized back in the third grade that when you were reading about photosynthesis, you were actually getting the basics of modern climate science. (*Photosynthesis* occurs when plants take in energy from the sun and carbon dioxide from the atmosphere and turn it into oxygen and sugars.) Figure 2-4 may jog your memory.

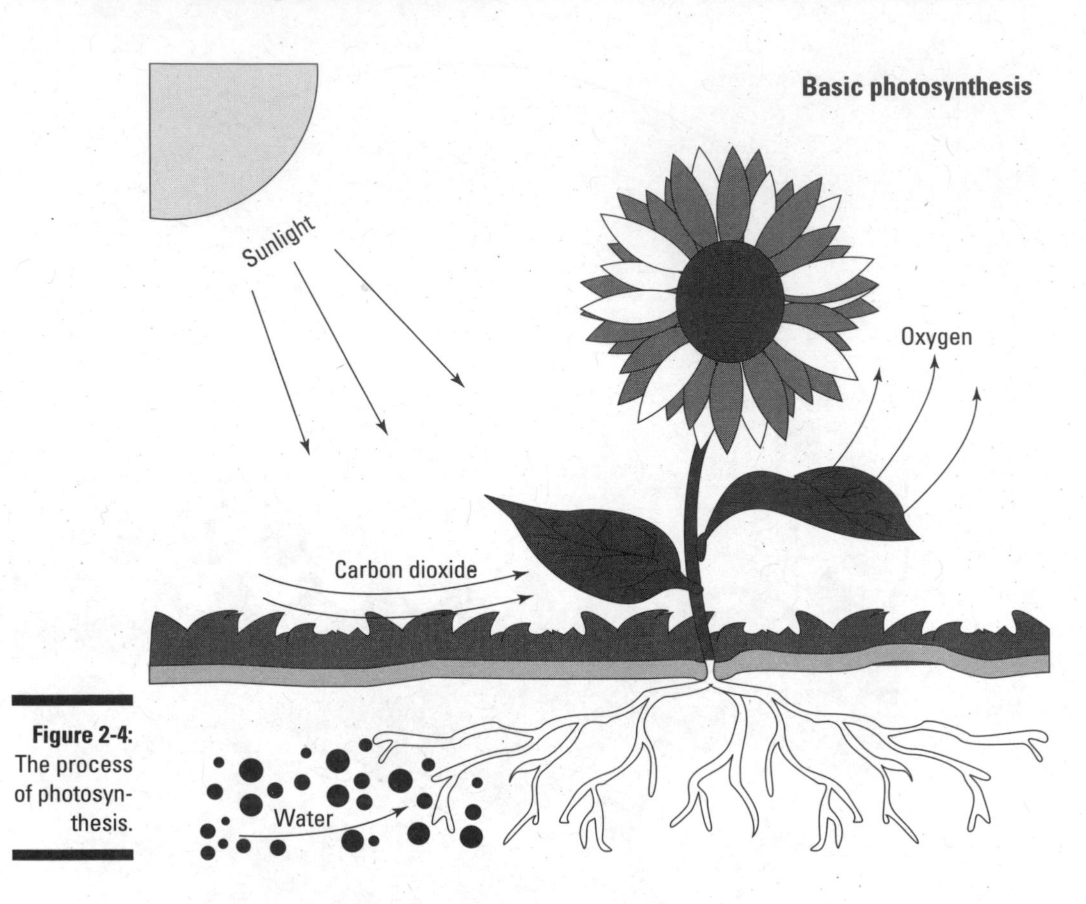

Figure 2-4: The process of photosynthesis.

Trees are our planet's biggest and most widespread plants. Our forests are wonderful carbon sinks.

The most effective carbon-trapping forests are tropical, such as those in Brazil and other South American countries. Most tropical forests are called rainforests (although not all rainforests are tropical). *Rainforests* grow in regions that get over 70.9 inches of rain each year. Because of all the rain they get, these dense, rich forests are full of biodiversity. And because of the tropical climate, which is always warm, these tropical forests work year-round. The tireless work that these trees do to sequester carbon is just one of the reasons to protect the tropical rainforests

Forests in Canada, the United States, and Russia aren't as effective at soaking up carbon because they take a rest in the winter but are still very important in the planet's carbon balance. The northern forests make up for the reality of their seasonal work, through the relatively richer and deeper soils. Northern forests store more carbon in carbon reservoirs, even though tropical forests take up more carbon on an annual basis.

Down to earth

Not just the trees and oceans store carbon; soil does, too. Plants draw in carbon dioxide and break it down into carbon, breathing the leftover oxygen into the atmosphere. The carbon makes its way into the soil through the plants' root systems or when the plant dies. See Figure 2-5 for a diagram showing how soil and trees exchange carbon dioxide with the air.

In this plant–soil relationship, most of the carbon is stored close to the top of the soil. Tilling the soil (mixing it up) exposes the carbon in the ground to the oxygen in the air, and these two elements immediately join to form carbon dioxide.

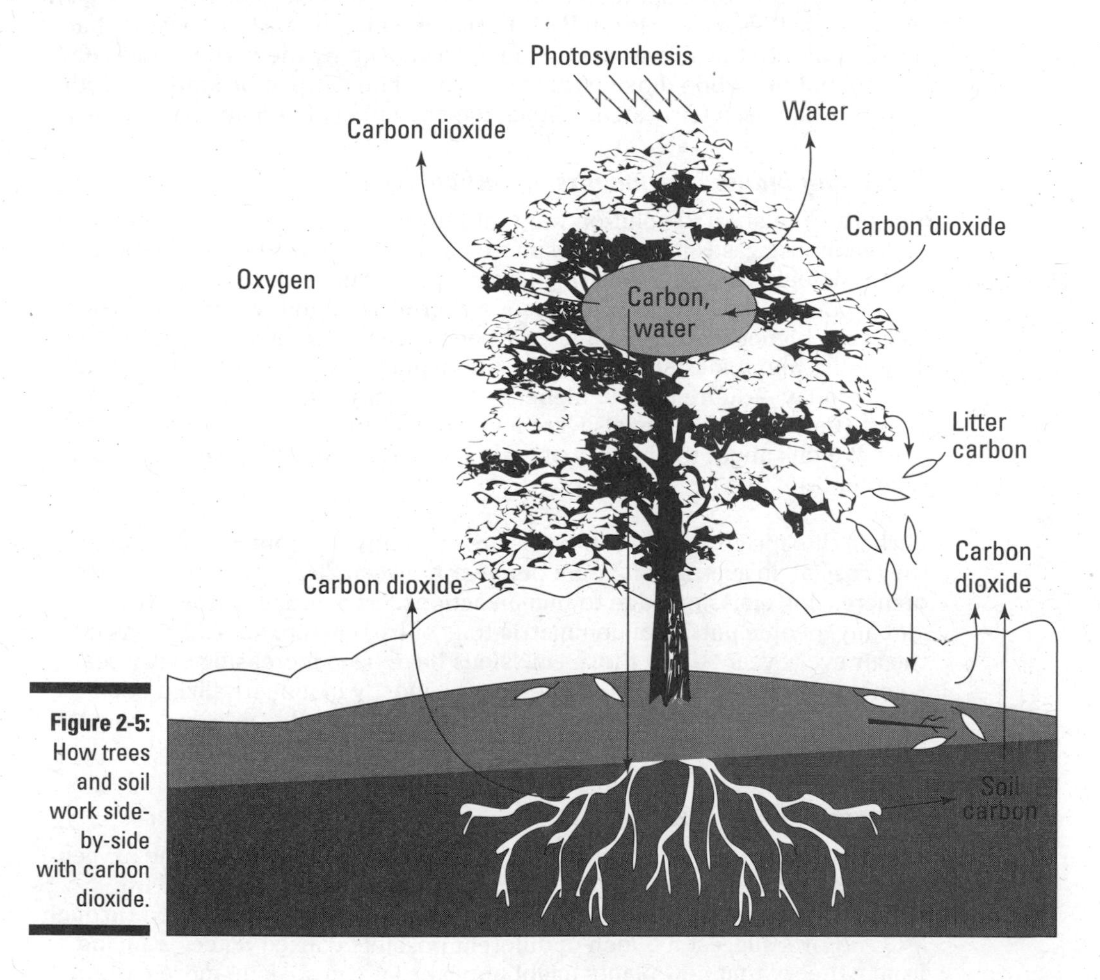

Figure 2-5: How trees and soil work side-by-side with carbon dioxide.

All together, vegetation and soil store about a billion metric tons of carbon every year, and another 1.6 billion metric tons move in and out between the land and the air. So far, the plants and soil have packed away 2,300 billion metric tons.

Investigating our impact on the carbon cycle

A lot of the carbon dioxide in our atmosphere is natural (you're breathing some out, right now), but human activities also contribute plenty of the gas (we discuss these activities in Part II). Historically, the carbon dioxide that people put into the air was pretty much soaked up by the carbon sinks, and the amount of carbon dioxide that was around before people started building factories had been fairly steady since the beginning of human civilization.

Producing industrial amounts of carbon dioxide

Since the Industrial Revolution went into full swing around 1850, the amount of greenhouse gases in the atmosphere has risen drastically. Due to burning fossil fuels, as well as clearing forests, people have almost doubled the carbon dioxide emissions in just over a century, and today, carbon dioxide levels are higher than they have ever been in recorded history (see Chapter 4 for more about fossil fuels). In fact, atmospheric carbon dioxide levels are higher today — more than $1/3$ higher — than at any time in the past 800,000 years. (Carbon dioxide levels were much higher millions of years ago, however. We talk about the history of carbon dioxide levels in greater detail in Chapter 3.)

Carbon dioxide concentration levels are currently at about 385 ppm, and they rose at an average of 2 ppm per year between 2000 and 2007 because of increasing emissions due to human actions. On average, in the 1980s, globally, people put 7.2 billion metric tons of carbon dioxide emissions into the air every year — and those emissions have been increasing every year. So many people are using so much energy, mostly in industrialized countries, that the amount of carbon that is being put into the air is knocking the carbon cycle off balance.

Plugging up the carbon sinks

The Earth's carbon sinks, which used to be able to handle everything oxygen-breathing creatures could throw at them, are not able to keep up with humanity's increased carbon dioxide production. Studies presented through IPCC reports suggest a bunch of different possible consequences, ranging from a theory that new plants might appear that can soak up more carbon dioxide to the idea that carbon sinks may become full and may no longer be able to absorb any more carbon dioxide. Like anyone who works overtime, carbon sinks could become weaker as they soak up more carbon dioxide. The ocean has stored carbon effectively in the past, but global warming might be causing the oceans to do just the opposite. The oceans have warmed 0.18 degrees Fahrenheit (0.1 degrees Celsius) on average in recent years, and carbon dioxide is less soluble in warm water. The oceans push the carbon dioxide that they can't dissolve into the air, instead. Data collected during the 1980s and 1990s suggested that both land and ocean sinks seemed to have kept up with growing emissions. However, more recent studies show that the carbon dioxide intake of some sinks, such as trees, is slowing down.

Sinks normally absorb about half of our emissions. So, if these sinks were to weaken, or even stop absorbing, they'd leave a lot more carbon dioxide in the atmosphere, on top of our already-increasing emissions.

Carbon dioxide is a greenhouse gas, which traps heat in the atmosphere. Higher levels of carbon dioxide are causing higher average temperatures.

Looking at the Other Greenhouse Gases

Carbon dioxide may get all the press, but 23 other greenhouse gases (in five main groups) also heat things up. Although they're present in much smaller amounts, these guys are actually far more potent, molecule for molecule, in terms of greenhouse effect. You might think of them as carbon dioxide on steroids. Table 2-1 shows you the power of some of these gases.

Because so many different types of greenhouse gases exist, people usually either talk about only carbon dioxide (because so much more of it exists than the others) or greenhouse gases in terms of *carbon dioxide equivalents* — how small an amount of the gas you'd have to put into the atmosphere to have the same warming impact as the current level of carbon dioxide. Referring to all greenhouse gases with this measurement makes assessing and measuring them that much easier. So, when we say greenhouse gas in this book, you can actually think of it as carbon dioxide equivalent emissions. No calculator needed.

Measuring in carbon equivalents means, for example, that 1 unit of methane equals 21 units of carbon. In other words, 1 metric ton of methane is just as bad as 21 metric tons of carbon dioxide. Thus, methane is 21 carbon dioxide equivalents, or 21 metric tons of carbon dioxide.

Svante Arrhenius: Early climate change scientist

Swedish chemist Svante Arrhenius was the first person to predict what the future atmosphere might look like in the wake of the Industrial Revolution.

He spent many of his days (and likely nights) at the end of the nineteenth century calculating how the carbon released by burning coal (the major source of fuel at the time) might actually change the atmospheric carbon balance. In the end, he calculated that humanity could double the concentration of atmospheric carbon — in 3,000 years.

The fact that Earth is now closing in on doubling that concentration just over 100 years after Arrhenius made his calculations has nothing to do with his grasp of chemistry or math — it has everything to do with the fact that he based estimates on what he knew.

The internal combustion engine was only a speculative invention, with none in use. No cars were on the road, and Arrhenius certainly had no idea about traffic jams, drive-through windows, or airplanes. Who could have imagined today's level of fossil-fuel consumption 100 years ago? After all, Arrhenius was a chemist, not Nostradamus!

Table 2-1 Global Warming Potential of Greenhouse Gases

Greenhouse Gas	*Global Warming Potential Over Time*	
	20 years	*100 years*
Carbon dioxide (CO_2)	1	1
Methane (CH_4)	56	21
Nitrous oxide (N_2O)	280	310
Hydrofluorocarbons (HFC) *Group of 13 gases*	3,327	2,531
Perfluorocarbons (PFC) *Group of 7 gases*	5,186	7,614
Sulfur Hexafluoride (SF_6)	16,300	23,900

Source: United Nations Framework Convention on Climate Change, GHG Data, Global Warming Potentials, `http://unfccc.int/ghg_data/items/3825.php`

Methane (CH_4)

Methane accounts for about 18.6 percent of the overall global warming effect from greenhouse gases, according to the World Meteorological Organization.

It's also 21 to 56 times more potent than carbon dioxide. Methane is to carbon dioxide what an espresso shot is to herbal tea. (See Table 2-1.)

Methane naturally occurs when organic materials, such as plant and animal wastes, break down in an *anaerobic* environment (an environment that contains no oxygen and includes the right mix of microbes and temperature). This breakdown creates methane, along with small amounts of other gases. The stomach of a cow, a landfill site, and a marsh are all prime examples of methane-producing environments.

Methane occurs naturally, but humans also add their fair share (okay, we're just being modest — humans add a lot). People contribute to the amount of methane when their garbage decomposes in landfills, and livestock contributes thanks to their flatulence (and maybe bad breath; expert opinion is mixed).

How methane gets into the atmosphere

Two-thirds of all the human-made methane comes from agriculture, and about half of that amount comes from rice crops. If you've ever seen rice being grown, you may remember that it's planted in a flooded field. Any dead organic matter falls to the bottom of the paddy, which is a perfect airless environment for creating methane.

All our cows, pigs, chickens, and other farm animals account for the rest. The food breaking down in their stomachs produces methane that they, shall we say, emit into the air — one way or another. All animals emit methane — yes, even you — but livestock's methane causes a problem because there are so many of these fairly big animals (with multiple stomachs — those would come in handy for extra slices of chocolate cheesecake . . .).

Humans also add methane to the atmosphere through treating wastewater and from landfills — all that garbage spews methane into the air while it breaks down.

People also use methane as a fuel. Natural gas is 90 to 95 percent methane, and when natural gas is extracted from the ground, some methane escapes into the air. (Read more about using natural gas as a fuel in Chapter 4.)

The mystery of stabilizing methane levels

Methane levels in the atmosphere seem to have stabilized at 1.8 parts per million, even though people keep dumping garbage, growing rice, and raising cattle. The experts haven't quite figured out why methane levels seem stable currently because methane production hasn't decreased noticeably, aside from a drop in rice crops in China.

Although methane level increases appear to have declined since the 1990s and have been nearly stable since then, this stability probably won't last. A lot of methane is frozen into the ground of the arctic, trapped by the permafrost.

Rising northern temperatures, however, are melting the soil. When the Arctic land thaws, it becomes swampland — and it starts dishing out methane like hotcakes. We look at this problem in greater detail in Chapter 7.

Nitrous Oxide (N_2O)

The amount of nitrous oxide in the atmosphere is even smaller than the amount of methane, but it accounts for about 6.2 percent of the overall greenhouse effect. The greenhouse effect of nitrous oxide (N_2O) per unit is almost 300 times more potent than that of carbon dioxide.

Unlike methane, this gas is actually still going up at a rate of 1 part per billion (ppb) each year — as of 2005, it was at 319 parts per billion.

Ocean- and soil-dwelling bacteria produce nitrous oxide naturally as a waste product. In agriculture, farmers encourage those natural bacteria to produce more of the gas through soil cultivation and the use of natural and artificial nitrogen fertilizers.

Dentists also use nitrous oxide as an anesthetic. (Laughing gas is nitrous oxide — not so funny now, is it? Though it's in such small amounts that you don't need to worry about your root canal adding to global warming.) Industrial processes (to create nylon, for example) produce nitrous oxide.

The biggest source of nitrous oxide, natural or human-made, is fertilizers used in agriculture. Fertilizers count for 60 percent of human-made sources and 40 percent of sources overall. Humans also add a lot of nitrous oxide to the atmosphere by using automobiles. Ironically, cars produce the gas as a side result of solving another environmental problem — see the sidebar "How good intentions increased nitrous oxide emissions," in this chapter, for the scoop.

Hexafluoro-what?

Hydrofluorocarbons. Perfluorocarbons. Sulfur hexafluoride. Try saying those names three times fast. They're as hard to say as they are effective at trapping heat. These three types of gases are all human-made and don't exist naturally in the atmosphere. They come from a number of different industrial processes that create air pollution.

Almost all car air-conditioning systems use the 13 hydrofluorocarbons (HFCs). (We look at how industry emits these greenhouse gases in detail in Chapter 5.) Most of the seven perfluorocarbons (PFCs) are byproducts of the aluminum industry. Sulfur hexafluoride (SF_6) comes from producing magnesium, and many types of industry use it in insulating major electrical equipment.

How good intentions increased nitrous oxide emissions

Fossil fuels (which we discuss in greater detail in Chapter 4) contain nitrogen. When cars burn gasoline, they give off the nitrogen-based chemicals nitrogen monoxide (NO) and nitrogen dioxide (NO_2) — together known as NO_x gases. These NO_x gases create acid rain and smog in cities.

In response to these environmental problems, governments in North America forced car companies to put catalytic converters in all their cars. Catalytic converters convert smog-causing chemicals into other chemicals that aren't as damaging to our lungs and don't cause acid rain.

Unfortunately, these catalytic converters turn NO_x gases, which don't have an effect on climate change, into nitrous oxide (N_2O), which does! (One more reason, if you need it, why everyone should drive less.)

Other players on the greenhouse gas bench

The two greenhouse gases that we talk about in the following sections do play a role in climate change, but they aren't on the United Nations list of 24 greenhouse gases and get left aside in most discussions about the impact of greenhouse gases on global warming — not for scientific reasons, but because of decisions made in international negotiations.

Water vapor

As we talk about earlier in this chapter, water vapor is a huge player in the greenhouse effect. As shocking as it may seem, good ol' H_2O (two parts hydrogen, one part water) causes the majority — 60 percent — of the planet's warming.

Unlike the production of the other greenhouse gases, humans don't directly cause the increase of water vapor. But the other gases that are produced heat up the atmosphere. When plants, soil, and water warm up, more water evaporates from their surfaces and ends up in the atmosphere as water vapor. A warmer atmosphere can absorb more moisture. The atmosphere will continue to absorb more moisture while temperatures continue to rise. See Figure 2-6.

Water vapor also differs greatly from other greenhouse gases because the atmosphere can hold only so much of it. If you've ever watched a weather forecast, you've heard the term *relative humidity,* which refers to the amount of water vapor currently in the atmosphere compared to how much the atmosphere can hold. On a really hot and sticky day, the relative humidity may be 90 percent — the atmosphere has just about taken in all the water vapor it can. When the relative humidity reaches 100 percent, clouds form, and then precipitation falls, releasing the water from the air.

Running up emissions with your sneakers

Some of the sources of these gases are really wild. Here's one: Nike came out with the popular Nike Air shoe, a running shoe with a cool little air-filled bubble in the heel, in the late '80s. That bubble was filled with — you guessed it — a greenhouse gas! (Sulfur hexafluoride, to be exact.)

The amount of greenhouse gas in those shoes all together added an equivalent of about 7 million metric tons of carbon dioxide — or the emissions from 1 million cars — into the air when they hit the garbage dump after being worn out. In the summer of 2006, after 14 years of research and pressure from environmental groups, Nike stopped using the greenhouse gas in shoes and replaced it with nitrogen. We're glad that bubble burst.

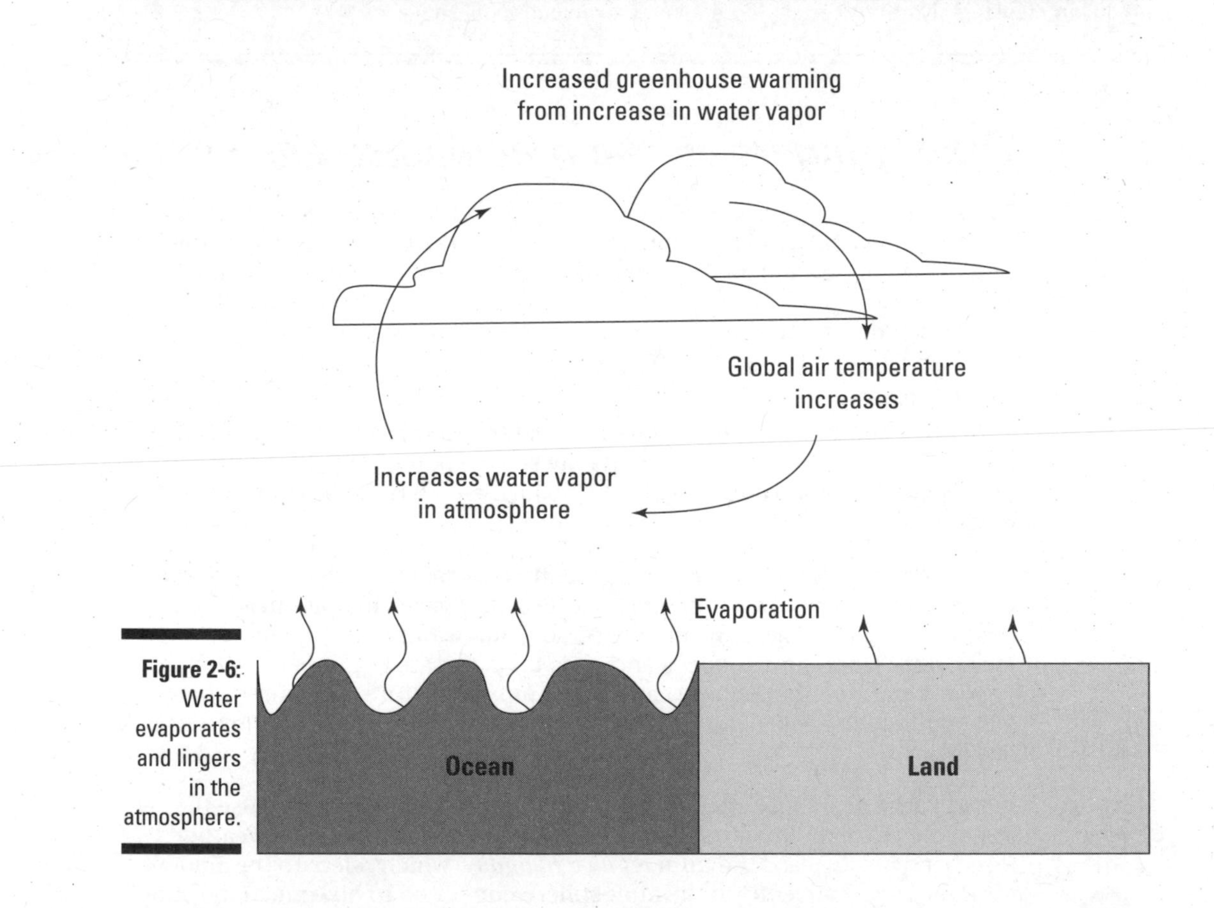

Figure 2-6: Water evaporates and lingers in the atmosphere.

Ozone depleters

Chlorofluorocarbons (CFCs) are also considered greenhouse gases, responsible for about 12 percent of the greenhouse effect the planet is experiencing today. You don't find much of these CFC gases around anymore because the Montreal Protocol of 1989 required countries to discontinue their use. CFCs break down the *ozone* — the layer throughout the stratosphere that intercepts the sun's most deadly rays. (Without the ozone layer, the sun's ultraviolet rays would kill all living things.) CFCs were mostly used in aerosol spray cans and the cooling liquids in fridges and air conditioning.

Because CFCs are already regulated under the Montreal Protocol, they're not regulated under the Kyoto Protocol, the agreement to reduce greenhouse gases that became active in 2005. (Read more about what gases are covered under the Kyoto Protocol in Chapter 11.)

Chapter 3

The Big Deal about Carbon

In This Chapter

- Looking into what else contributes to global warming
- Linking carbon dioxide to temperature trends
- Understanding what happens when the temperature becomes too hot to handle
- Limiting greenhouse gas emissions

The greenhouse gases that are emitted when humans burn fossil fuels are causing global warming. No other theory explains the climate changes that scientists are observing.

That's right — global warming is a theory. A scientific *theory* is based on a set of principles that describe a particular phenomenon — like the theory of gravity. Theories aren't technically facts, but sometimes theories become so strong that people accept them as facts. But how do you know this theory is correct? Can you really trust all those bigwig scientists? And if it's correct, what does this theory suggest is going to happen next? We answer those questions in this chapter.

Considering Causes of Global Warming Other than Greenhouse Gases

Sure, some uncertainty exists around how much of the planet's warming is due to natural effects and how much of it is due to human activity. When it comes to the culprits behind climate change, greenhouse gases, although important, aren't the only players.

The climate is an incredibly complex system affected by the sun, cloud cover, and complex long-term trends. Because of this complexity, various theories in science — other than the main one about greenhouse gas emissions — may at least partially account for global warming.

The Intergovernmental Panel on Climate Change (IPCC) reports that the idea that climate change is being caused by natural changes alone has a 5-percent chance of being true, and that the idea that increasing carbon dioxide levels due to human activity are at least partly behind climate change has a 95-percent chance of being true. So, although some uncertainty exists, the debate is largely settled.

Solar cycles

The sun has different cycles, and the Earth's climate changes over time in response to these cycles. The sun goes through *irradiance cycles,* which is when the amount of solar radiation reaching the Earth varies. Scientists only found out about these cycles recently. Two of these cycles seem to exist, one running for 11 years and the other running for 22 years. Scientists don't yet know whether the sun goes through other, longer irradiation cycles — these unknown cycles might, in part, cause climate change.

The solar cycles do affect climate in the short term, but the U.S. National Oceanic and Atmospheric Administration (NOAA) reports that the impact from the light intensity of the sun versus the impact from greenhouse gas emissions is a ratio of approximately 9 to 40. So, greenhouse gases have more than four times the effect of solar cycles.

Other cycles that concern the sun are the Milankovitch cycles. Although they sound like Mr. Milankovitch's bike collection, *Milankovitch cycles* are actually natural cycles of the Earth — one of these cycles, for example, is the way in which the planet tilts towards or away from the sun. These cycles may explain the *glacial cycles* — the ins and outs of ice ages. (We talk about the Milankovitch cycles in more detail in the section "Making the Case for Carbon," later in this chapter.)

Although they're very important, the Milankovitch cycles have minimal effect on climate, in comparison to the effects from greenhouse gas emissions, when you look at them in terms of relatively short timescales — from decades to centuries. Overall, the IPCC says that the sun likely has little to do with global average temperature rises since 1950. In fact, models suggest that the Earth would be cooling if not for increases in greenhouse gases.

Cloud cover

Scientists have known for a long time that climate affects rainfall. NASA (the National Aeronautics and Space Administration in the U.S.) has shown that the relationship may work in reverse, however — that the changing rain patterns might, in turn, indirectly affect global warming. Rainfall patterns correspond to cloud cover. Depending on their thickness and shape, clouds can reflect light during the day and hold in surface heat overnight. (See Figure 3-1.) The amount of water vapor in the air has recently increased (which we talk about in Chapter 2), which means more clouds, which means more rainfall. This increase in cloud cover might help explain why nighttime temperatures are rising more than daytime temperatures in global warming trends. It gives a whole new meaning to having a hot night!

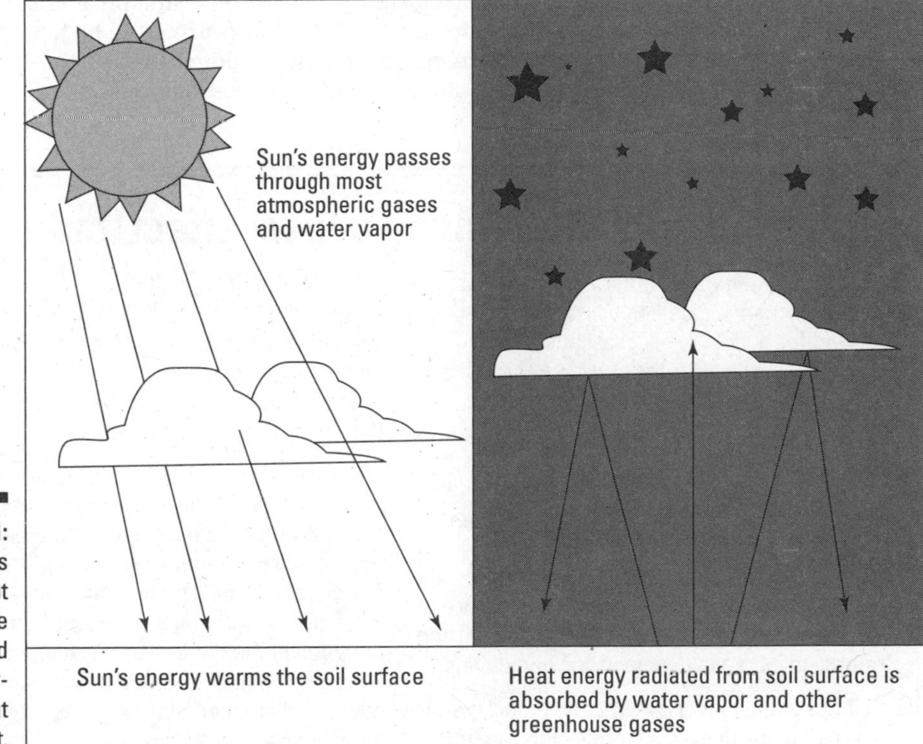

Figure 3-1: Clouds reflect light during the day and hold in surface heat overnight.

Ultimately, however, increased cloud cover seems to be a result, not a cause, of climate change. But, like the increased water vapor, it may further aggravate global warming.

Long-term climate trends

Over the course of many millions of years, the Earth's temperature has varied widely. The latest change in temperature may be entirely in keeping with that variation — but it's unlikely. Scientists know that the current period of ice ages started about 2 million years ago, and since 800,000 years ago, the planet started into a cycle of an ice age every 100,000 years or so. Currently, the Earth is in an *interglacial period* — meaning the weather is warm and stable enough that humans can develop and expand societies. Human civilization started at the beginning of this interglacial period about 10,000 years ago. Given that past warm interglacial periods lasted about 10,000 years, on average, scientists would expect the world to be getting cooler, not warmer. In fact, it appears that this cooling was happening between the middle ages and the 19th century (the little ice age), but then came the Industrial Revolution.

El Niño: Global warming cause, effect, or both?

El Niño is a natural weather cycle that has the power to change global temperatures. It's been around for hundreds — possibly millions — of years. El Niño involves the tropical Pacific Ocean warming by 0.9 degrees Fahrenheit (0.5 degrees Celsius) or more for about three months at a time. This warmed water eventually loses that heat to the atmosphere, causing the average air temperature (at the surface, or where humanity lives in the lower part of the atmosphere) to go up a few months later, which then alters the overall climate temporarily. The temperature of the ocean then settles back down to normal and returns to its regular cycle of ups and downs.

Scientists don't yet know whether global warming is affecting these cycles, but global warming and El Niño cycles are very interrelated. Part of the reason scientists can't distinguish between the impacts of global warming and El Niño on climate is because they're so linked, and they both influence many different aspects of regional climate, that they can actually change one another. A computer model giving future scenarios of climate that includes both El Niño and global warming doesn't yet exist because of the difficulty that exists between identifying the separations between the two.

Some models say El Niño will become stronger, but others say it'll weaken. Evidence suggests that El Niño cycles have been stronger and happening more often over the past few decades, and climate models project that climate change will cause sea-surface temperatures to rise in the tropical Pacific Ocean — similar to El Niño conditions. Scientists are working constantly to advance their understanding of the relationship between climate change and El Niño.

Two unknown questions remain: How much temperature rise is the result of El Niño and how much is the result of global warming? Are El Niño temperatures any higher because of global warming?

Making the Case for Carbon

Scientists have collected evidence that points to the build-up of carbon dioxide in the Earth's atmosphere as the most likely cause of climate change.

They can measure exactly how much carbon dioxide has been in the Earth's atmosphere historically. *Climatologists* (scientists specializing in climate science) have drilled deep — as deep as 2 miles (3 kilometers) — into ancient ice in places such as Antarctica and Greenland. They've pulled up *ice cores* — long, thin samples of many layers of ice that has been packed down over thousands of years, which look like really (really) long pool noodles. (Figure 3-2 shows two scientists drilling for an ice core sample.) When the layers are clearly visible, the ice core looks like a pool noodle with horizontal stripes.

Scientists can date an ice core by counting the layers of ice — just like you can tell the age of a tree by counting its rings. The layers of ice tell them exactly when the ice was formed. Each layer of ice includes little pockets of trapped air. These frozen air bubbles are like time capsules of the ancient atmosphere. They're full of gas, including carbon dioxide, that has been trapped for hundreds of thousands of years. Each layer of ice in the ice core also contains *deuterium,* a hydrogen isotope that enables scientists to determine what the temperature was when that ice layer was formed.

An atmospheric temperature change of just 1.8 degrees Fahrenheit (1 degree Celsius) leads to a change of nine parts per million (ppm) in the amount of deuterium stored in the ice. By contrasting the ancient temperatures revealed through the analysis of the layer's deuterium and carbon dioxide, scientists can glimpse the relationship between historical levels of carbon dioxide and temperature. The two run side by side almost like the lanes of a race track.

Scientists still don't know the exact cause and effect relationship between greenhouse gases and temperature throughout the planet's history. The cause of the last ice age, for instance, probably wasn't a drop in atmospheric carbon dioxide, but a result of the Earth tilting away from the sun in a phase in the planet's Milankovitch cycle (which we discuss in the section "Considering Causes of Global Warming Other Than Greenhouse Gases," earlier in this chapter). This cooling then spurred the atmosphere's carbon dioxide to drop, and the two events in tandem brought about the ice age. Ultimately, scientists still aren't sure whether temperature affects carbon dioxide, or whether carbon dioxide affects temperature — it's a question of which came first, the chicken or the egg.

Figure 3-2: Drilling for an ice core sample.

National Oceanic and Atmospheric Administration

What scientists do know for certain is that a distinct pattern and relationship between carbon dioxide and temperature exists; when one is high, so is the other, and when one is low, the other plunges, too. Scientists also know that the Milankovitch cycle has little to do with climate change over the past 200 or 300 years. In that time, human-produced carbon dioxide levels have skyrocketed, and temperature is starting to follow. As a result, scientists are certain that human-produced greenhouse gases are currently warming the Earth. This close relationship between greenhouse gas concentrations and temperature suggests these higher levels of carbon dioxide will cause temperatures to continue rising.

Figure 3-3 shows the historic connection between carbon dioxide concentrations and fluctuations in temperature, as captured in ice-core deuterium levels.

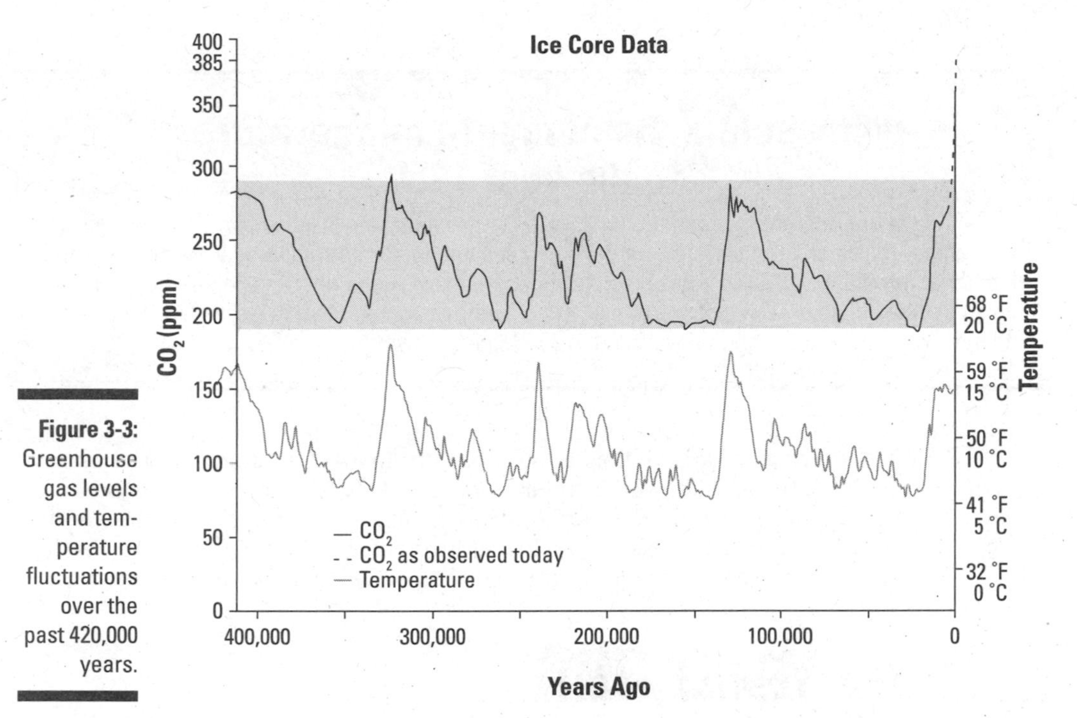

Figure 3-3: Greenhouse gas levels and temperature fluctuations over the past 420,000 years.

Data: National Oceanic and Atmospheric Association (NOAA) Vostok Ice Core data and Mauna Loa CO_2 observations. Graph: John Streicker

The Consequences of Continued Carbon Dioxide Increases

If scientists are right about the connection between carbon dioxide and climate change, then what comes next? The past is all very interesting, but it's history. What the future holds concerns all of humanity — and the predictions that scientists have for the future are alarming.

As the ice cores demonstrate, carbon dioxide levels have always fluctuated (check out "Making the Case for Carbon" earlier in the chapter for more information), but the atmosphere now has 35 percent more carbon dioxide than at any time in the last 800,000 years. Historically, carbon dioxide has reached highs of 280 parts per million (ppm) at a maximum. The atmosphere is now at 385 ppm.

Why scientists compare temperatures to the year 1850

The international scientific community uses the temperatures at the time just before the Industrial Revolution (1850) as a baseline. They do so because human contributions to climate change were not significant before that time. By measuring the build-up of greenhouse gases and temperatures compared to what they were before the Industrial Revolution, they're measuring the impact that is largely attributable to human activity.

This increase in carbon dioxide is an extraordinary shift. If present trends continue, the Earth's average temperature is likely to increase by 3.6 to 10.8 degrees Fahrenheit (2 to 6 degrees Celsius) above 1850 temperatures — and that temperature increase could be disastrous for all life on Earth. The Earth's temperature has already risen approximately 1.4 degrees F (0.8 degrees C).

The tipping point

The *tipping point* is the point at which something has gone too far — or past the point of no return. Think of slowly going up the first climb on a roller coaster. After you go over the top, no one can stop the ride.

Scientists believe that climate change has a tipping point, when the damage becomes too great to be reversed. After this point, not only can nothing reverse the impact on the planet, but little could stop that impact from increasing, either — it just keeps getting worse.

To determine the climate's tipping point, scientists first had to look at what would happen if, say, temperatures went up by, say, 3.6 or even 10.8 degrees Fahrenheit (2 or 6 degrees Celsius) above 1850 levels. (These temperature increases refer to the global average, which we discuss in the following section.) To figure out the effects of temperature increases, scientists depend on sophisticated models. Not models that you build when you're a kid by using papier mâché — these kinds of *models* are mathematical, designed to be run on a computer, and simulate the functioning of the Earth's atmosphere and climate. (See the "How climate models work" sidebar, in this chapter, for more information.) Researchers input data about the climate and how it works, and then start modifying that data to create various alternative scenarios.

How climate models work

The climate is affected by both the atmosphere (the part that everyone talks about the most) and the oceans. Changes in the air happen quickly, and changes in the oceans happen very slowly. So, scientists have been able to study air changes relatively easily, but they have quite literally had to wait and see what happens to the oceans. And because the ocean actually affects the bulk of the climate, they're also having to wait and see what happens to the entire climate. So, scientists need climate models, projected scenarios created by super computers, to help predict major climate changes.

The most complex climate models, such as those used at NASA, look at the Earth in three dimensions. The computer divides the atmosphere and oceans into square columns. Each of these columns has its own set of weather information based on the history and current status of the area. This information gives the computer a base to work from. Then, the researcher running the model changes the numbers to see what would happen if one condition changed, such as air temperature. For short-term projections (looking forward a day to a month), an advanced computer can make the calculation in 20 minutes. But making longer-term projections (such as 50 years from now) can take a month or two. A global circulation climate model can take as long as a year to produce results after researchers input all the variables.

Scientists figured out, for example, how hot the climate would need to become to melt the entire world's ice sheets — this melt would cause sea levels to rise, which would flood coastal cities around the world. At the same time, the scientists figured out the amount of greenhouse gases needed to reach these temperatures.

By looking at these different possibilities, scientists could tell which effects of climate change humans can deal with and which ones are beyond humanity's ability to adapt to or control.

The IPCC defined an average global temperature rise of about 3.6 degrees Fahrenheit (2 degrees Celsius) above 1850 levels as the official climate change warning zone, but that temperature increase is below the point of no return, or the tipping point. If the temperature goes to 5.4 degrees Fahrenheit (3 degrees Celsius), 7.2 degrees Fahrenheit (4 degrees Celsius) becomes inevitable. At 7.2 degrees Fahrenheit (4 degrees Celsius), 9 degrees Fahrenheit (5 degrees Celsius) becomes inevitable. And so on. The increases soon outstrip any human ability to slow or control those increases. The increased warming becomes inevitable because of positive feedback loops (see Chapter 7 for the lowdown on feedback loops). Melting permafrost releases methane, speeding more warming. Melting icecaps reveal more dark water, speeding warmer ocean temperatures and more ice melt. Dryer conditions lead to more forest fires, releasing more carbon and causing more warming. This domino effect could lead to an unlivable world.

No one knows exactly where that tipping point for global warming is. Scientists know only that humanity has a chance to avoid it by holding carbon dioxide concentrations to no more than 450 ppm, to keep the planet's average temperature increase at or below 3.6 degrees Fahrenheit (2 degrees Celsius). The IPCC says that the average global temperature will rise by 3.6 degrees Fahrenheit (2 degrees Celsius) when total carbon dioxide levels reach 425 to 450 ppm. Trouble is, it's moving upwards at 2 ppm per year and is currently at about 385 ppm of carbon dioxide as we type this.

Uncertainty always exists when it comes to making predictions. Acknowledging this, the IPCC actually hedged its bets slightly and said that the warning zone is between 2.7 and 4.5 degrees Fahrenheit (1.5 and 2.5 degrees Celsius) above 1850 levels — or 1.3 and 3.1 degrees Fahrenheit (0.7 and 1.7 degrees Celsius) above current levels. Its assessment report suggested that people consider the critical warning zone to be 2.7 degrees Fahrenheit (1.5 degrees Celsius) above 1850 levels — or 1.3 degrees F (0.7 degrees C) above current levels — on the basis that it's better to be safe than sorry. (The United Nations calls this philosophy the *precautionary principle.*)

In fact, humans have already committed the planet to a 2.9-degree Fahrenheit (1.6-degree Celsius) temperature rise versus 1850 levels — or 1.4 degrees Fahrenheit (0.8 degree Celsius) beyond today's temperature — by the end of the century, based on current climate models. Humanity could level off this increase if it implements the necessary solutions by 2010.

A few degrees is a lot

Three or four degrees Fahrenheit seems like a small number to make a big deal about. You might even be thinking that an extra 3.6 degrees Fahrenheit (2 degrees Celsius) seems like a perfect amount of global warming. Your garden would grow better, you'd be hitting the beach more often, and the golf season might be longer, right? But 3.6 degrees Fahrenheit (2 degrees Celsius) is actually a lot. The IPCC reports that the global average temperature in the middle of the last ice age was only 7.2 to 12.6 degrees Fahrenheit (4 to 7 degrees Celsius) colder than it is today.

This increase of 3.6 degrees Fahrenheit refers to the average global temperature, but average numbers hide the extremes on either end. For example, you can dive into a pool that has an average depth of 1 foot (30 centimeters) if it's 10 feet (3 meters) at the deep end. Right now, the average global temperature is 60 degrees Fahrenheit (15.6 degrees Celsius). Of course, temperatures can be much colder than that in the winter and way warmer in the summer. In that same 60-degree Fahrenheit (15.6-degree Celsius) global average, you can go skiing in the Alps or swimming in the Caribbean.

The rate of warming in Arctic regions is twice the global average, whereas regions close to the equator could see very little change. Because of the Arctic's high warming rate, less sea ice is left at the end of each summer — in fact, the area covered by perennial sea ice has gone from covering 50 to 60 percent of the Arctic region down to covering 30 percent, according to data processed by NASA.

What happens when the mercury rises

A climb of 3.6 degrees Fahrenheit (2 degrees Celsius) in the world's average temperature may trigger a number of unpleasant consequences:

- Increased droughts in what are now semi-arid areas
- Rapid Arctic ice melt
- Fast loss of *permafrost,* the frozen ground in the Arctic
- Increased damage to cities and major infrastructures because of higher-intensity storms and floods
- A possible 30-percent increase in species extinctions

For more on the consequences of global warming, check out Part III.

Figure 3-4 outlines the changes that different temperature increases, up to and beyond 3.6 degrees Fahrenheit (2 degrees Celsius), may bring.

Cutting Back on Carbon

Modern civilization probably won't stop producing greenhouse gas emissions altogether. But to stop levels from passing much farther beyond 450 ppm — and thus limit the world to a 3.6-degrees Fahrenheit (2-degrees Celsius) temperature rise — more people need to reduce their emissions. Because of this global need, the first step of the Kyoto Protocol is very important — it requires just over a 5-percent reduction (below 1990 levels of emissions) by 2012. (See Chapter 11 for more about international climate change agreements.)

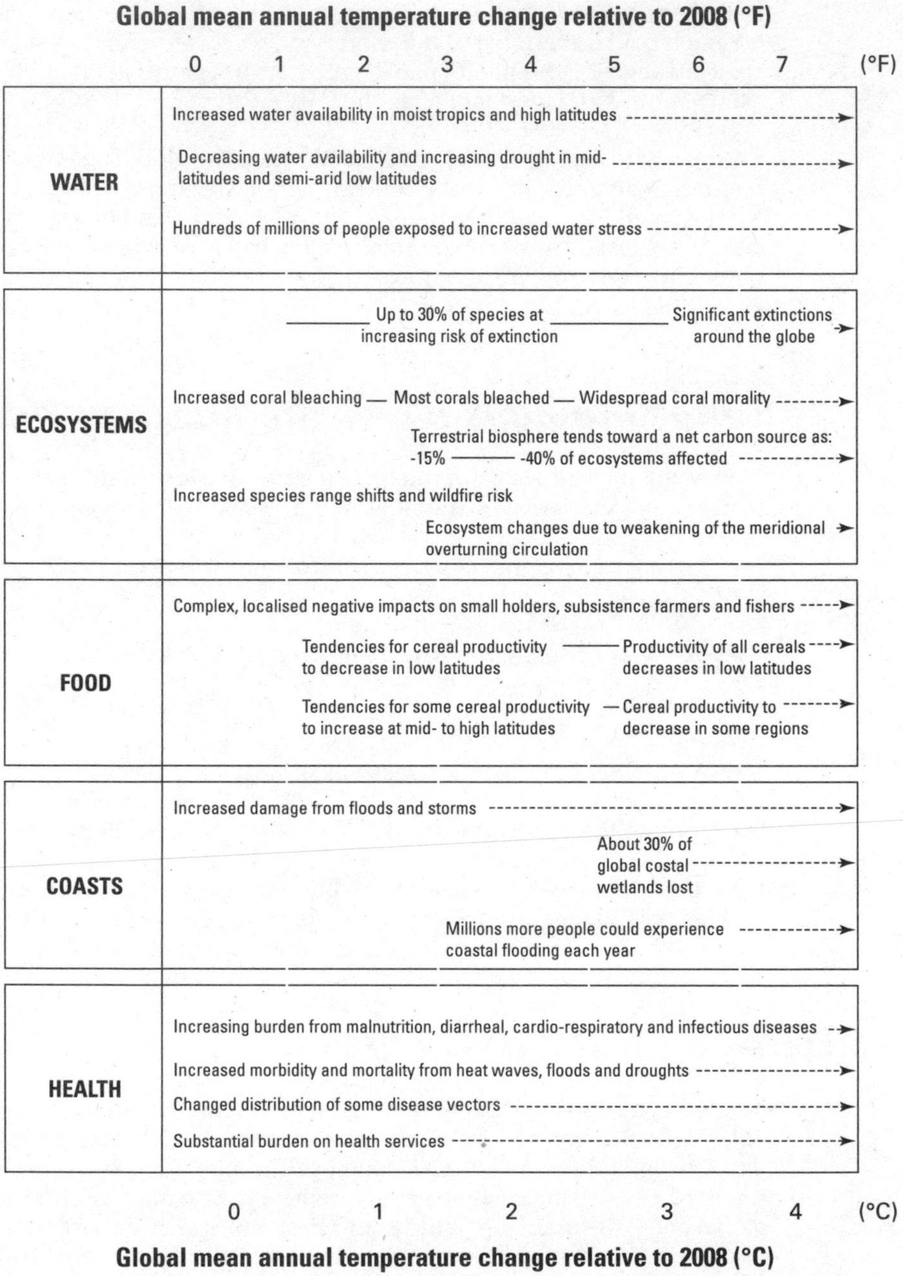

Global mean annual temperature change relative to 2008 (°F)

0 1 2 3 4 5 6 7 (°F)

WATER	Increased water availability in moist tropics and high latitudes ---> Decreasing water availability and increasing drought in mid-latitudes and semi-arid low latitudes ---> Hundreds of millions of people exposed to increased water stress --->
ECOSYSTEMS	Up to 30% of species at increasing risk of extinction — Significant extinctions around the globe → Increased coral bleaching — Most corals bleached — Widespread coral morality ---> Terrestrial biosphere tends toward a net carbon source as: -15% — -40% of ecosystems affected ---> Increased species range shifts and wildfire risk Ecosystem changes due to weakening of the meridional overturning circulation →
FOOD	Complex, localised negative impacts on small holders, subsistence farmers and fishers ---> Tendencies for cereal productivity to decrease in low latitudes — Productivity of all cereals decreases in low latitudes ---> Tendencies for some cereal productivity to increase at mid- to high latitudes — Cereal productivity to decrease in some regions --->
COASTS	Increased damage from floods and storms ---> About 30% of global costal wetlands lost ---> Millions more people could experience coastal flooding each year --->
HEALTH	Increasing burden from malnutrition, diarrheal, cardio-respiratory and infectious diseases ---> Increased morbidity and mortality from heat waves, floods and droughts ---> Changed distribution of some disease vectors ---> Substantial burden on health services --->

0 1 2 3 4 (°C)

Global mean annual temperature change relative to 2008 (°C)

Figure 3-4: Effects from climate change will intensify as temperatures rise.

Modified and based on Figure 3.6. Climate Change 2007: Synthesis Report. Fourth Assessment Report. *IPCC. Cambridge University Press.*

Some people say that the world will have a hard time reducing emissions by even 5 percent because of the energy-intensive ways that people currently live in industrialized countries. Others argue that this emissions reduction may be humanity's best choice because it's the safest route to take. (We consider ways that governments can help lower emissions in Part IV, and in Part V, we look at how businesses and individuals can cut back on emissions.)

Countries haven't yet agreed to any greenhouse gas reduction targets past the year 2012, but they did set in motion new negotiations in November 2007 aimed at deciding these future greenhouse gas reduction targets. The IPCC recommends reducing carbon emissions by 50 to 80 percent below 1990 levels by 2050. Many countries and groups have committed to an aggressive 2050 goal independently, including the European Union, California, and the World Mayors Council. (See Chapter 10 for solutions being implemented by governments around the world.) A few countries, such as Canada, won't commit to the IPCC's recommendation until they know for sure that they can hit the target.

The main reason a major reduction in greenhouse gas emissions is so important is because a chance exists that the planet's climate situation could get worse than predicted. For example, changes could speed up the warming cycle because of positive feedback loops (see Chapter 7 to find out about these feedback loops). Or parts of the carbon cycle could weaken because of increasing temperatures, meaning that not as much carbon would be sucked up by carbon sinks such as forests and oceans — leaving humans to deal with more emissions than expected. (The precautionary principle, which we talk about in the section "The tipping point," earlier in this chapter, looks more appealing by the day!)

If civilization keeps doing what it's doing, even with no increase in how much greenhouse gases it produces from year to year, it'll lock Earth into the 3.6-degrees Fahrenheit (2-degrees Celsius) deal by the end of 2010, though humanity wouldn't see the temperature shift until years later. Every time a person drives a car, for example, he or she releases carbon dioxide that will act as a force for global warming for the next 100 years!

The global climate system has long lag times. The atmosphere doesn't turn on a dime. The damage humanity does today will have an effect over a century. If people keep on with business as usual, the world might be committing to the 3.6 degrees Fahrenheit (2-degrees Celsius) rise sooner than expected.

Part II
Tracking Down the Causes

"Sure, global warming's affected us. But my garden's never been so lush!"

In this part . . .

What's causing climate change? Many factors are playing a role, but one substantial source of greenhouse gases stands out: fossil fuels. We dig deep, exploring where these fuels come from and why they're causing so much trouble. We also look at how major industries, from manufacturing to logging to farming, are contributing to climate change, and we investigate how everyone is unwittingly contributing to the problem.

Chapter 4

Living in the Dark Ages of Fossil Fuels

In This Chapter

- Investigating where our energy comes from
- Recognizing the differences between coal, oil, and natural gas
- Connecting the dots between population growth, economic expansion, and climate change

Wherever you live, however you heat your home, and however you get around, you probably meet most of your energy needs by burning fossil fuels, such as coal, oil, and natural gas.

Burning these fossil fuels releases large amounts of greenhouse gases (we talk about those gases in Chapter 2). In fact, just over two-thirds of human-produced greenhouse gases in the atmosphere come directly from burning fossil fuels. In this chapter, we examine the types of fossil fuels, look at how people use them in their day-to-day lives, and assess fossil fuels' overall contribution to climate change.

From Fossils to Fuel

A lot of people know that fossil fuels pollute and produce carbon, but they don't understand why. To understand that, you need to know where fossil fuels, such as coal, oil, and natural gas, come from. They're literally derived from fossils of past living matter.

Talk about fossils, and the first things that may come to mind are dinosaurs. But when it comes to the fossils in fossil fuels, they're actually fossils from *before* the time of the dinosaurs — starting off as decomposing plant material (not decomposing dinosaurs).

Many of these plants grew in swamps that used to cover even the northernmost parts of the globe 300 to 400 million years ago. Usually, plants and trees rot away into the soil, but swamps don't have enough air (it's what scientists call an *anaerobic environment*) for the usual decomposition process to happen. Instead, over time, these dead plants and trees sank to the bottom of the swamps where they eventually turned into peat. The peat was buried and compressed under layers of sediments such as sand and silt. As these sediments turned into rock, more pressure was piled on the peat below it. The moisture was squeezed out of the peat like water squeezed out of a sponge, turning the peat to fossil fuels. So, millions of years later, fossil fuels are typically found deep underground.

Similar fossil fuels are also found under the ocean, where sea plants and old shells were buried and pressed down under the ocean sediment. See Figure 4-1 to get an idea of what this process looked like.

Not all plant matter in those ancient swamps and in the oceans turned into fossil fuels. The process needed the right conditions, such as enough pressure and the correct bacteria. Although many of these plants were very different from anything known about today, they sucked up carbon dioxide from the atmosphere and gave off oxygen, like all plants still do (check out Chapter 2 for more about photosynthesis).

When fossil fuels are burned for energy, those fuels release the carbon, in the form of carbon dioxide and other gases that these ancient plants stored. (For more on why releasing carbon dioxide into the atmosphere is problematic, refer to Chapter 2.)

Greenhouse gases are released not only when these fuels are burned, but also when they're retrieved from the earth. Extracting the fuels and processing them into their final forms requires fossil fuels, and thus produces carbon dioxide. The oil has to be taken out of the earth, transported to a refinery, processed into a usable form, and transported to its final destination. Because traditional sources of oil have begun to dry up, industry has turned to sources such as Alberta, Canada's tar sands, which require even more energy to yield any fuel. (See the sidebar "How much oil is left" for more about the possible end of oil and the sidebar "Athabasca tar sands: A sticky situation" for information about the tar sands.)

Fossil fuels give a one-time-only burst of energy. Take them out of the ground and burn them, and that's it. The supply of fossil fuels is limited, and after people use them up, civilization will have to wait millions of years before any more exist. That's why, no matter what, civilization will have to rely on a diversity of fuel sources to produce energy in the future. We talk about alternative energy sources in Chapter 13.

Figure 4-1: How fossil fuel is created.

Examining the Different Types of Fossil Fuels

Coal, oil, and natural gas are all fossil fuels, but they're not all the same. They differ in how they're used, how much they're used, the greenhouse gases that they release when they're burned, and even where they come from.

When land plants, such as trees, decomposed hundreds of millions of years ago, they pressed together into a solid form known as coal. Plants and animals in the oceans decomposed in a similar way — sinking to the bottom of the ocean, getting buried under sediments, forming peat, and eventually being compressed into fossil fuels such as oil.

Each type of fossil fuel has a different amount of carbon in it, so it puts a different amount of carbon dioxide into the air when it's burned. Coal releases the most carbon dioxide when burned, natural gas the least. In the following sections, we take a closer look at the different types of fossil fuels, starting with the worst offender, coal, and working our way down to natural gas.

Coal

You may think that coal was king forever ago, but about a quarter of the world's energy still comes from this fossil fuel. Coal was the first fossil fuel that humans burned for energy — in fact, the use of coal predates written history.

Coal is a very dirty fuel. Because it's essentially carbon, it releases carbon dioxide when burned, along with many other dangerous pollutants. In December 1952, to cite one dreadful example, a massive lull in air circulation trapped the coal smoke from tens of thousands of London homes over the English city, creating a blanket of pollution. In four days, the deadly smog (the name comes from combining smoke and fog) killed upwards of 4,000 people directly, with 8,000 more succumbing to respiratory illnesses later on.

Some of the noxious stuff inside coal includes sulfur dioxide, mercury, and a huge array of polyaromatic hydrocarbons (cancer-causing and hormone-disrupting toxic chemicals, also in oil and gas). And it doesn't stop there. Coal also releases arsenic and cyanide; *carcinogens* (things that promote cancer), such as benzene, naphthalene, and toluene; and a witch's brew of other nasties.

Clean coal doesn't exist. Options for cleaner uses of coal, however, do.

The first step taken to reduce pollution from coal plants was aimed at reducing nitrous oxide emissions. To do this, the coal is burned at incredibly high temperatures (around 1500 degrees F) — this is considered a "low temperature" compared to the 2500 degrees F at which coal is usually burned. At such a low temperature, nitrogen does not combine with oxygen, thus no nitrogen oxide (NOx) is created. This process happens during the burning process and reduces many pollutants but does nothing to reduce carbon dioxide.

The next step in pollution reduction was when coal-fired power plants in many industrialized countries added scrubbers in the 1970s to capture the sulfur and prevent it from falling to Earth as acid rain. Scrubbers are technically called *flue gas desulfurization units* — devices installed right in the flue. The device sprays a specially made liquid mix of water and powdered limestone right into the emissions coming from the burning coal. The spray immediately soaks up and becomes one with the sulfur, trapping it in this new solid material.

Another way of "cleaning" coal is called *fluidized bed combustion,* where the coal actually becomes liquid in the bed of the furnace. Scrubbers and fluidized bed combustion reduce emissions of nitrogen and sulfur dioxide, but not of carbon dioxide. Industries were even able to extract the sulfur and sell it, increasing their profits. Removing sulfur dioxide was a step in the right direction for solving the problem of acid rain, but these scrubbers, again, do nothing to reduce carbon dioxide, mercury, or the whole array of other pollutants.

Research and development teams are devoting a lot of time and energy to producing a type of coal that doesn't add to greenhouse gases. One idea suggests turning coal into a gas and stripping the carbon dioxide out of that gas, then storing the carbon dioxide in the ground. The technology to actually strip the carbon dioxide from the gas doesn't yet exist, but the carbon-storage technology does (and parts of Europe already use it). Until the day that carbon dioxide can be stripped out of coal, conservation practices and replacing coal-fired power plants with cleaner, renewable fuels are the most effective and sustainable ways to reduce greenhouse gases. (Flip to Chapter 13 for more on clean fuels and carbon storage.)

Oil

Today, oil provides about 40 percent of the world's energy. People use it each time they fill up their cars, get on a plane, or turn on an oil furnace. Oil is also the key raw material used for manufacturing a wide variety of very common products, including plastics (from food containers to toys), artificial fibers, and a host of other goods such as hair gels, shampoo, deodorant, and dishwashing liquid.

All petroleum products start out as crude oil. A barrel of crude is more than just a barrel of crude! You can have sweet crude and regular (or sour) crude. (Sweet crude has lower sulfur content.) Then, the light and heavy crude classifications depend on, quite literally, how light and heavy the crude is. Whatever the type, crude oil is the straight-up oil, before anyone does anything to it. Refineries process the crude oil to make gasoline for cars, diesel fuel for trains and trucks, heating oil for homes, and jet fuel used in airplanes.

How much oil is left?

Climate change activists have been urging people to reduce their fossil fuel consumption because of the impact on global warming, but another compelling reason to cut back on oil use exists: It's running out. There are roughly 1.4 trillion barrels of *proved oil reserves* left, according to British Petroleum reports. Proved oil reserves are estimated volumes of oil with an 80 to 90 percent certainty, according to the International Energy Agency. The argument about when, exactly, the world will run out of fossil fuels (particularly oil) has been going on for decades. When the Club of Rome released its famous Limits to Growth report back in 1970, it said with certainty that the planet was running out of oil. The 1990s once again saw the rise of the argument that the planet would soon be out of oil. This time, the alarm was raised by geologist M. King Hubbert's concept of peak oil. Hubbert's *peak* referred to the point at which people would begin depleting known reserves, or when oil consumption is higher than oil production. Today's peakers, as they're known, are finding a lot of evidence that this point has passed. They argue not that Earth is running out of oil (which it will eventually, without question), but that Earth has already run out of cheap oil.

Cheap oil has been the lifeblood of the post–World War II economic boom. The July 18, 2007, report that the U.S. National Petroleum Council (NPC) sent to the U.S. Secretary of Energy states that 80 percent of today's oil production must be replaced with new sources of oil or other energy sources within the next 25 years. That's a daunting prospect. It may or may not be possible — but even attempting it will surely be very expensive.

The International Energy Agency estimates that the world economy needs to find an additional 3.2 million barrels of oil a day. Every single day, the world's petroleum geologists, and oil and gas companies, must find new sources of oil — new oil fields and new bitumen deposits equal to 3.2 million barrels of oil — just to keep the current supply steady.

Peakers argue that when the crunch hits, it will really hurt, causing recessions. People won't be able to afford the gas to fuel their cars, and suburbs will suffer. But people can see it coming and can start investing in energy efficiency and smarter ways of using the oil that's available. And using more renewable fuels could help cushion the blow of more expensive, dwindling fossil fuels. Climate change activists know that the atmosphere is running out of space for the wastes from burning fossil fuels, no matter how much longer supplies last.

Virtually everyone agrees: The age of cheap oil is over.

When you think of oil, you probably imagine it shooting up out of the ground like a fountain. But those *Beverly Hillbillies* days of "black gold, Texas tea" are long over. Humanity has already used up most of those easy-to-tap reserves.

Oil is starting to play hard-to-get. Although many disagree about whether humanity has hit *peak oil* — the point of maximum production of oil, after which the supply begins to be depleted (see the sidebar "How much oil is left?" in this chapter, for details). Companies are now discovering about only 30 billion new barrels a year, in comparison to the 200 billion barrels a year that they found in the early 1960s.

Oil is found in harder-to-reach places these days, and sometimes, companies need large amounts of water to push the oil out. Look at the Athabasca tar sands in Alberta, Canada, for example. The oil industry used to consider separating the oil from this thick, gooey mixture of clay, sand, water, and oil too expensive. But now that the price of oil is so high, the industry decided that the process of physically pressing the oil out of the sand is worth the cost. (Check out the sidebar "Athabasca tar sands: A sticky situation," in this chapter, for an in-depth look at this process.)

Offshore reserves have long been a source of oil — they're still under the ground, but also under the water. You can find these reserves off the east coast of Canada, in the Gulf of Mexico, and off the coast of Norway. Now, however, the search for new offshore oil fields is heading for more remote and fragile areas, such as the Beaufort and Chukchi Seas. These two diverse ocean ecosystems host thriving wildlife, on which the local indigenous peoples depend.

On the ground, both government and oil companies are proposing oil drilling in protected areas such as the Arctic National Wildlife Refuge. Companies are also proposing projects in the Amazon rainforest in Ecuador, another fragile ecosystem that's also a vital part of the planet's carbon cycle (refer to Chapter 2 for more information).

No one would have considered these sources of oil a decade or so ago. But dwindling oil supplies and rising prices have changed all that. The world economy has become used to oil higher than $120 a barrel. And to think that the world was shocked when oil hit $30 a barrel in the 1970s!

Natural gas

Natural gas is mostly methane, which makes it a little different than the other fossil fuels (check out Chapter 2 for more about methane). The cleanest of all fossil fuels, natural gas gives off only carbon dioxide and water when it burns. Rotting trees and plant matter release methane if the conditions are wet and airless. Natural gas can usually be found around coal beds or oil fields. Although this underground natural gas will run out someday, humanity can produce pure methane in other ways, such as capturing the methane gas that comes from the rotting waste in landfills.

Athabasca tar sands: A sticky situation

No matter what the name suggests, the *tar sands* (also called *oil sands*) aren't a tarry version of the Sahara Desert. They're *boreal* forests (coniferous forests, found between 50 and 60 degrees North, across northern Canada, Russia, Alaska, and Asia, as well as Scandinavian Europe) and *muskeg* (a type of wetland found in boreal and arctic areas) that cover a sandy soil that's 10-percent *bitumen,* a viscous material that resembles tarry molasses. To get the oil, you have to squeeze this bitumen out of the sands.

The Athabasca tar sands hold 173 billion barrels of proven retrievable oil, and up to 315 billion barrels of potentially retrievable oil if new technologies are developed. These reserves make it the second-largest oil patch in the world, after Saudi Arabia. Reaching the bitumen involves stripping away the muskeg and boreal forest — a single mine may need more than 6,500 hectares (16,000 acres) of forest cleared.

After removing the muskeg and forest, oil companies dig the bitumen out of open-pit mines that are 245 feet (75 meters) deep. To extract the bitumen that lies even deeper, they have to pump huge amounts of water and steam into the ground to loosen it up and bring it to the surface. They use between 2.5 and 4 barrels of water for every barrel of oil extracted, depending on how deep the bitumen lies. This process creates a lot of wastewater, full of toxic waste, that they store behind enormous dikes.

The Alberta government permits the tar sands operation to use more than 1,177 cubic feet (359 million cubic meters) of water annually — twice as much as the city of Calgary, which has a population of over a million people, uses in the same period. A 2006 report by the Canadian National Energy Board questioned whether the project's massive water use was sustainable. Many towns and communities, such as Fort McMurray, also rely on the river from which this water is drawn for their drinking water — the water that the mining uses may one day seriously stress the water source of Fort McMurray residents.

Heating up the bitumen and extracting it from the tar sands takes a lot of energy — energy that's supplied by . . . burning fossil fuels. These operations use the equivalent of a third to a half a barrel of oil for every barrel of oil produced. (Anyone see a losing cycle here?)

So, the mining industry consumes a huge amount of energy in order to produce oil, which primarily the United States buys for cars that don't have proper energy efficiency standards (California excluded!).

Canada's decision to keep expanding and developing the tar sands is an example for other nations of what not to do — while making oil development a top priority it's impossible for Canada to decrease its greenhouse gas emissions. The report compared it to the American decision to encourage coal as a form of energy independence and to Brazil's clearing of rainforests.

Natural gas is almost pure methane by the time it reaches your doorstep. A quarter of the world's energy comes from natural gas. The Intergovernmental Panel on Climate Change (IPCC) reports show that the Earth contains more natural gas than regular oil, but that natural gas is patchy and spread out in comparison to oil, making it harder to tap into.

Because of how relatively clean it is, some energy analysts have promoted natural gas as a clean-energy fuel to replace coal in power plants. But this solution may not be all it's cracked up to be. Natural gas is

- **A finite resource:** Supplies of natural gas are tight. High-end estimates predict that world supplies will run out in 50 years at the current rate of use, and in as few as 5 years in North America.
- **Difficult to transport:** Moving natural gas involves liquefying it first, which requires a lot of energy. This liquefying process also creates carbon dioxide emissions, depending on the source of energy. (For instance, coal-fueled energy would create more emissions than hydroelectric energy.)
- **Potentially dangerous:** Concerns exist around possible pipeline explosions, as well as the environmental damage created by gas exploration. Leaks and explosions do happen: On December 14, 2005, the community in Bergenfield, New Jersey, awoke to a tremendous explosion caused by a leaking natural gas pipeline that demolished an apartment building and claimed three lives.

Fuelling Civilization's Growth: Adding to the Greenhouse Effect

Fossil fuels have been powering human development for a long time. Since the Industrial Revolution, civilization has been steadily consuming more fossil fuels; consequently, more and more carbon dioxide has been pumped into the atmosphere.

The world's growing population has been a key factor in the increasing levels of greenhouse gases in the atmosphere. Earth's population was 1.2 billion in 1850, when the Industrial Revolution was taking place. In the past 50 years alone, the population has doubled from 3 billion to more than 6 billion — today, the population is 6.5 billion. Even if the per capita use of fossil fuels had remained relatively stable, the amount of greenhouse gases would have increased. And, of course, use keeps on growing.

Luckily, population growth is slowing and should level off. (The bad news is that this isn't expected to happen until the Earth's population reaches 9 billion people). Nevertheless, the estimate of population numbers leveling at 9 billion is a better outcome than some growth curves that put us at exponential growth to over 12 billion. It all depends on reducing fertility rates, which all depends on improving the economic, educational, and political status of women and girls.

Countries don't produce carbon dioxide emissions equally. Unfortunately, North Americans are over-achievers when it comes to creating carbon dioxide emissions. According to the World Resources Institute, one North American emits the same amount as two and a half Europeans, 10 Bangladeshis, or more than 20 sub-Saharan Africans!

Population pressure is a factor, but a growing economy also plays a large role in boosting emissions of fossil fuels. The modern world economy has been hard-wired to use them. Businesses and governments used to think that economic growth depended on using more and more fossil fuels. But then, in the 1970s, when major members of the Organization of Petroleum Exporting Countries (OPEC) drastically reduced oil exports for political reasons, oil prices jumped. As a result, governments encouraged people to use less oil — so they drove less, bought fuel-efficient cars, and practiced energy conservation. Industrialized nations took the first, tentative steps in reducing the use of fossil fuels.

But, after the mid-1980s (when oil prices dived), some old addictions took over. In the U.S., for instance, the size of the average home (which needs fossil fuels to heat it) has increased by 50 percent since 1970 (though the size of the average family has decreased), and more drivers are using large, fuel-guzzling vehicles, such as SUVs. (You can read about improving home energy use in Chapter 18, and about more fuel-efficient vehicles in Chapter 17.) Countries such as Iceland and Sweden, however, switched to a renewable energy base and stayed that way.

Historically, the stronger a country's economy, the more greenhouse gases it produces. A strengthening economy means a greater consumption of fossil fuels — just look at the rapid growth of the auto industry in China, which promises to surpass the U.S. in production and sales.

But even as the economies of developing countries grow, they still emit only a small portion of what people in industrialized countries do, per capita. They have a lot of catching up to do. When we wrote the first draft of this chapter in 2007, the world's biggest greenhouse gas polluter as a whole country was the United States. But, since then, China's emissions of greenhouse gases have already surpassed the U.S.'s. The total pollution from developing countries is expected to exceed the pollution from the industrialized world by 2030. (We take a look at developing nations in Chapter 12.)

A low-carbon future is possible. In fact, the IPCC says countries need to move quickly to clean energy or else the course of climate change will become irreversible. It recommends that governments establish effective policies that support clean energies and wean the world off oil. We talk more about government solutions in Chapter 10, and explore energy alternatives in Chapter 13.

Some countries show that economic growth and carbon dioxide emissions aren't necessarily intertwined. Sweden has seen 44-percent economic growth while reducing its greenhouse gas emissions to 8 percent below 1990 levels. Sweden has pledged to become the first country on Earth to go off oil by the year 2020.

Chapter 5

Getting Right to the Source: The Big Emitters

In This Chapter

- Understanding the different ways energy creates emissions
- Seeing how transportation adds to greenhouse gases
- Looking at how fewer trees and more farms mean more warming

No one likes the blame game; pointing fingers and making accusations doesn't solve anything. When it comes to global warming, no one person, industry, or country is responsible for the build-up of greenhouse gases. Nobody wakes up in the morning and decides to try to make global warming worse. But every activity adds up.

This chapter zeros in on where the bulk of emissions comes from: the big greenhouse gas emitters, including power producers, buildings' energy systems, industry, shipping goods, agriculture practices, and deforestation. We get into further detail in Chapter 6, looking at how individual decisions play into the bigger picture.

Power to the People: Energy Use

In an automated age, just about everything in our civilization requires power — from the furnaces in our buildings to the batteries in our MP3 players to the engines in the trucks on the highway. Unfortunately, that power isn't always Earth-friendly; most of it comes from fossil fuels (which we talk about in Chapter 4). Energy use accounts for about two-thirds of human-caused greenhouse gas emissions in the world.

About half of all the world's energy-linked emissions come from the Group of Eight (G8) countries: Canada, France, Germany, Italy, Japan, Russia, the United Kingdom, and the United States. The world's remaining 186 countries account for the rest. Sounds unbalanced, but it looks like developing countries are moving quickly to tip the scales. Rates of energy use are growing fastest in developing countries.

Producing electricity

Generating electricity produces large amounts of greenhouse gases. Large-scale power plants are incredibly inefficient, and they essentially waste as much as two-thirds of the fuel that they use, either as heat sent up smoke stacks or through electricity lost along transmission lines.

Power plants take one kind of energy and turn it into another, electricity. Frequently, that initial energy source is a fossil fuel. Coal or oil plants, for example, burn the coal or oil to produce enough heat to boil water to generate steam. The force of the steam turns turbines, which creates mechanical energy. This process generates electricity that's delivered to buildings. Check out Figure 5-1 to see how a coal-powered electricity plant works.

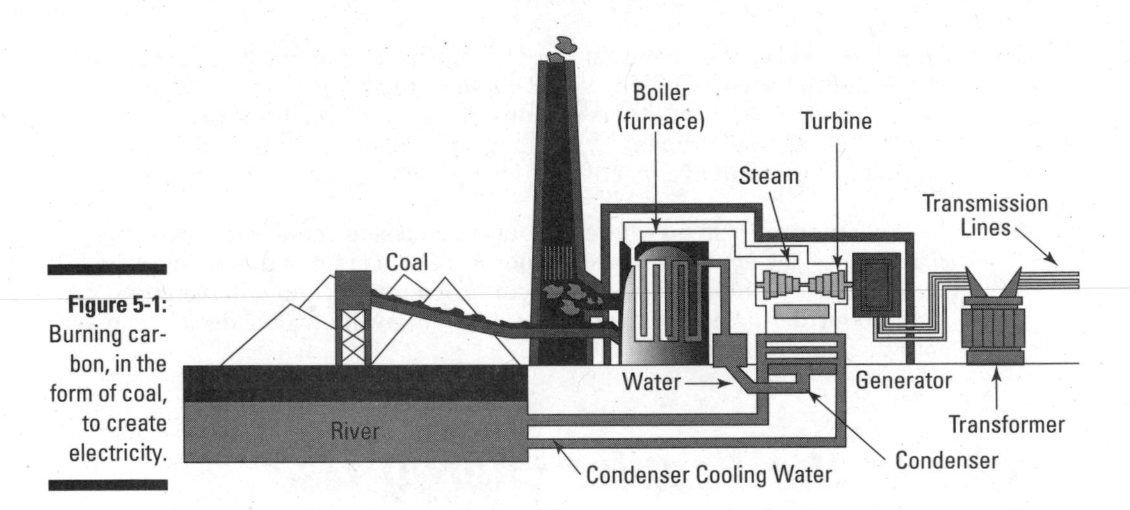

Figure 5-1: Burning carbon, in the form of coal, to create electricity.

The International Energy Agency (IEA) reports that 80 percent of Australia's electricity comes from coal. The U.S. uses coal for 50 percent of its electricity. Canada, on the other hand, is taking advantage of its surroundings and generates 58 percent of its electricity from hydropower, meaning far fewer emissions are generated to create electricity.

Incredibly, many countries around the world are building more coal plants, despite the rise in awareness about climate change. The IEA says that coal demand is growing 2.2 percent each year and probably won't let up for at least the next 20 years. New coal plants keep popping up, even in the European

Union (EU) and North America. Why? Coal is cheap. Even with potential carbon taxes (which we discuss in Chapter 10), coal is *still* the cheapest source of energy in the world because the infrastructure is already in place, and the price of coal is low. The downside is that each new plant represents a 30-year commitment to infrastructure supporting burning fossil fuels and to continuing greenhouse gas emissions. (Refer to Chapter 4 for more about coal and the myth of clean coal.)

Our dirty old coal plants are just that — very dirty and very old. They rely on a technology that we should have left behind in the 19th century. We have more efficient forms of energy that don't need fossil fuels at all and go through far fewer steps — such as hydropower, which uses the run of the river to turn the turbines. In Chapter 13, we look at all the other ways that people can produce electricity without boiling water. (You can still make a cup of tea by using electricity flowing from a non-polluting wind turbine or solar photovoltaic unit, though.)

Using up energy in buildings

About 15 percent of civilization's emissions come from the energy expended to heat and produce electricity for buildings, according to the International Energy Agency (IEA). Two-thirds of that energy is used in homes (see Chapter 6); commercial and institutional buildings, such as schools, colleges and universities, hospitals, shopping malls, and office buildings, use the other third.

The power plants in a region or oil-run heating systems right in a building make heating and electricity possible. These plants and systems burn fossil fuels and create the greenhouse gas emissions.

Think of all the ways that people use — and waste — energy and electricity in buildings. How many times have you walked into an office building in the middle of summer and felt like you were being transported back to the Ice Age because of the extreme air conditioning? People wear suit jackets and sweaters indoors in the summer, and shirt sleeves and skirts in the winter, reversing the seasonal shifts!

Many other factors influence the greenhouse gas-producing intensity of buildings, such as whether they

- Are well-insulated (see Figure 5-2 to see where an inadequately insulated building's heating and cooling escapes, wasting energy and creating more emissions)
- Have proper caulking around doors and windows
- Make maximum use of daylight to avoid relying on electric lighting

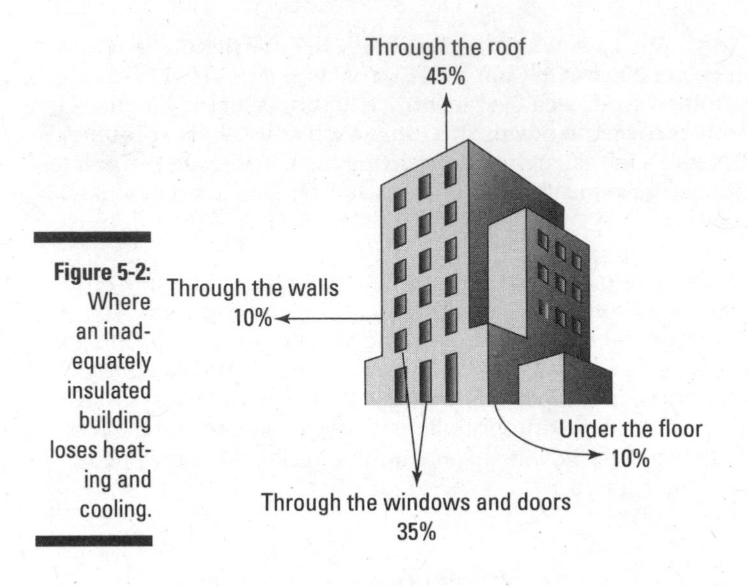

Figure 5-2: Where an inadequately insulated building loses heating and cooling.

In Chapter 18, we talk about how you can reduce the amount of energy required to heat or cool your home.

Powering industry

When people look for a single place to lay the blame for global warming, they often talk about "industry." After all, it's a pretty broad target, including just about any business that manufactures. And, admittedly, industry is responsible for a big chunk of human-produced greenhouse gases. Just below 40 percent of global carbon dioxide emissions come from energy used by industry. Industry almost exclusively releases the rest of the human-produced greenhouse gases — including methane, sulfur hexafluoride, perfluorocarbons (PFCs), some hydrofluorocarbons (HFCs), and some nitrous dioxide. (We talk about greenhouse gases in detail in Chapter 2.)

Industry produces greenhouse gas emissions in two main ways:

- Burning fossil fuels to create energy and electricity.
- Plain emissions that come directly from materials. The ingredients in cement production, for example, give off carbon dioxide, nitrous oxides, and sulfur dioxide.

Over half of the emissions from big industry — from coal-fired power plants, other oil and gas activities, construction, manufacturing, electricity production, pulp and paper plants, and so on — come from developing countries and countries that have economies in transition. About a third of that energy still comes from old coal-burning electricity plants. The smaller industries in developing countries often don't have the financial means or the technology to move beyond the older and inefficient equipment. Industrialized countries, on the other hand, can advance quickly in developing and using new technologies that have higher energy efficiency. (We talk about outdated equipment and other issues surrounding developing countries in Chapter 12.)

The products of industry that require the most energy are

- Metals (primarily iron and steel)
- Mineral production (including cement)
- Oil
- Pulp and paper

Iron and steel

Steel is the metal that we use most in the world. Steel-making is responsible for up to 7 percent of global human-created carbon dioxide emissions. Economists have long associated steel production with advanced industrial economies. A quarter of all steel today is produced in China.

Steel is manufactured in three ways; unfortunately, the most common method produces the most carbon dioxide emissions. Most steel is made by melting down iron ore in coke-fueled furnaces. (No, the furnaces aren't guzzling soda — *coke* is made by baking coal in airless ovens at extremely high temperatures.) Carbon and other elements are then added to the iron to produce steel. This common way to make steel uses the most energy. You might have heard of crude steel, which, despite its name, is far more ecologically polite. It's made from the scraps from the regular steel-making process, which are melted down again in electric furnaces. This process uses about a third of the energy that the regular process does. The last steel-making process doesn't actually make steel at all, but makes a stand-in: direct reduced iron. Natural gas is used to melt down the iron, a production process that creates half the carbon dioxide emissions of mainstream production. (Industries in industrialized countries are trying to increase the use of this practice to modernize, with the benefit of reducing carbon dioxide emissions.)

Most steel-production emissions come from producing coke and burning coke and coal in furnaces for energy.

Other metals

Producing other metals also uses a lot of energy, although people make a much smaller volume of other metals, compared to steel. Despite the smaller amount of metals produced, the industry's emissions include some of the worst greenhouse gases, such as perfluorocarbons (PFCs), from manufacturing aluminum, and sulfur hexafluoride (SF_6), from producing magnesium. Just one ton of SF_6, for example, has just as powerful a global warming impact as 16,000 metric tons of carbon dioxide.

Aluminum production has one of the highest rates of energy use of all industries. It requires tremendous amounts of electricity to convert raw materials to finished aluminum. Take a look at a can of soda. If the aluminum in the can came from *virgin materials* — in other words, if none of the aluminum in that can came from recycled aluminum — the energy required to produce that can is equivalent to 8 ounces of gasoline. (Recycling aluminum isn't just about saving space in the landfill; it's about saving energy!)

Some leading aluminum companies, such as Alcan, have been striving to reduce emissions. Alcan has reduced its emissions by 30 percent since 1990, even though its production has jumped 50 percent. Getting rid of PFCs was the key. (Flip to Chapter 14 for the scoop on industries that are implementing greenhouse gas-saving solutions.)

Oil

Even though we usually talk about oil as a fuel, we sometimes forget that it's first and foremost a product. The oil industry burns fossil fuels and creates emissions when drilling for and extracting fossil fuels! We go into all the fossil fuel details, including energy-intensive oil sand developments, in Chapter 4.

Pulp and paper

Mills that create pulp and paper products use a lot of energy. Like for so many other industries and our own homes, a power plant often creates that energy (we discuss how generating electrical power contributes to greenhouse gases in the section "Producing electricity," earlier in this chapter).

Pulp and paper plants have made improvements by using wood waste to generate power and by capturing waste heat. In Canada, many pulp and paper plants generate their own electricity by using hydropower.

Pulp and paper mills also produce emissions in the paper-making processes. Mills' wastewater releases methane, and the mills produce solid waste when manufacturing the pulp and paper. And, if you want to trace the pulp and paper industry even farther back, the carbon sink of trees is lost when the mills harvest those trees for paper products. (Don't feel guilty. The book you're holding is made from recycled paper and is as environmentally friendly as Wiley could make it!)

The Road to Ruin: Transportation and Greenhouse Gases

Many products today originate in countries that have cheaper labor — you'll likely be hard-pressed to find products made in your own country most of the time. The computer that Zoë used to write this book was made in China. A banana that you ate for breakfast might be from Ecuador, and your favorite t-shirt may have come from Bangladesh.

Cheap goods at a high price to the climate

Today, both raw materials and finished products must travel great distances, often thousands of miles, before goods end up in the hands of consumers. And because these goods typically travel by ship, truck, or airplane (all big emitters of greenhouse gases), most of the products that we buy and food that we eat come with a hefty climate change price tag attached.

"Big Box" stores, common throughout North America, offer low-priced merchandise that comes with a high carbon cost. These retail outlets are located where land-based tax rates are lower, and space is cheaper than in downtown areas. Such areas aren't well serviced by mass transit, meaning that shoppers must drive to the stores, causing greater greenhouse gas emissions. With shelves often stocked with goods that have been manufactured in places where environmental protections are weak — and then shipped and trucked long distances — the deals they offer aren't nearly as sweet when you factor in the climate cost of greenhouse gases.

Keep the air pollution on your side of the border

Trucks idling add serious amounts of air pollution and greenhouse gases to the atmosphere.

The North American Commission for Environmental Cooperation (a body established alongside the North American Free Trade Agreement to review environmental impacts in Canada, the U.S., and Mexico) found that mortality among Mexican children in towns along the U.S.-Mexico border had gone way up as a result of the pollution from diesel trucks idling at border crossings.

Air quality has also gone down and respiratory illness up in southwestern Ontario because of the extra time security checks take at the U.S.-Canada border since the terrorist attacks on September 11, 2001.

Holding ______ accountable for air travel emissions

If you can fill in the blank, you can solve a global problem.

Air travel is tricky. It's increasing rapidly, and there are more planes in the sky every year. But international air travel was left outside the Kyoto Protocol. The negotiators couldn't agree on which country should be responsible for emissions — the country from which the airplane took off or the one where it landed. International flights tally up to over 350 million metric tons of carbon dioxide annually, according to the International Energy Agency.

Government officials at international meetings still can't figure out which countries to hold accountable for the emissions that result from shipping goods internationally, which is a total of over 450 metric tons of carbon dioxide annually. The European Union, however, is hoping to introduce legislation that will put emissions from air travel between EU airports under the EU emissions trading market by 2011.

Keep on truckin'

Trucks have taken a larger and larger role in shipping goods in the last decade or so. And they're common — as you can tell by all the trucker stops, brake-check stations, and runaway lanes that accommodate them on highways. Trucks are everywhere on U.S. highways, so it's no surprise that the International Energy Agency (IEA) has ranked the U.S. as the highest-producing country of carbon dioxide emissions from domestic transportation — the U.S. creates over a third of the world's transportation emissions, a total of 1,800 million metric tons. The second-biggest producer of emissions from transportation (for now) is China, with a much lower 250 million metric tons.

Business used to move goods by railroad in large amounts and store the goods in warehouses. The advent of just-in-time delivery in manufacturing has increased the amount of greenhouse gases released per product. Rather than have warehouses well-stocked with inventory that kept manufacturers and retailers ready for their work and customers, just-in-time delivery moved goods to the highways, constantly bringing goods in at the moment they're needed. Fewer warehouses, more trucks. No wonder the greenhouse gas emissions from transportation rose!

Manufacturers essentially warehouse their goods on the highways in tractor-trailers. This approach wastes a lot of energy. For example, trucks that ship frozen goods really gobble up energy. Even if the trucker stops for the night, that truck has to keep running to prevent its goods from thawing.

Companies adopt just-in-time delivery because of the money it saves them; they don't need to spend money warehousing goods. Although moving goods by truck increases congestion on highways, air pollution, traffic accidents, and (of course) greenhouse gas emissions, the companies don't have to pay for these negative side effects.

If someone could persuade manufacturers to go back to shipping goods by rail, those manufacturers would reduce the amount of greenhouse gases they create. If countries modernized the rail freight system, using safer, more efficient tracks and cleaner-running engines, they'd reduce those greenhouse gases even more. They'd still need trucks for getting products from the trains to all the stores or suppliers within a region, but they could reduce the practice of long-haul trucking.

Draining Our Carbon Sinks: Land Use

Talking about how land use relates to giving off greenhouse gas emissions might seem a little odd, at first. After all, you don't see corn crops or clear-cut land puffing out greenhouse gas emissions like a factory smoke stack. The link between greenhouse gas emissions and land has to do with how plants and soil work together.

Plants and soil are major carbon sinks that soak up carbon dioxide from the atmosphere. They take carbon dioxide out of the air and store it every day without us even noticing.

But when people cut down forests or over-till soil, carbon dioxide is released, not absorbed. In fact, in the last 50 years, the carbon dioxide emissions connected to land use have doubled from 1 to 2 billion metric tons per year, according to the Intergovernmental Panel on Climate Change (IPCC). In the following sections, we go into exactly why land use has increased carbon dioxide emissions so drastically.

Timber! Deforestation

Destroying forests is a major human cause of increased carbon dioxide concentrations. The IPCC reports that deforestation accounts for between 25 and 30 percent of human greenhouse gas emission additions to the atmosphere, when trees are cleared for timber, farming, and land development. It stresses that the largest problems are deforestation in tropical regions and the ability for forests to re-grow in temperate regions.

Trees are carbon *sinks* — they suck carbon dioxide out of the air, hold onto the carbon, and release the oxygen back into the atmosphere. The fewer the trees, the less carbon dioxide gets drawn out of the air. Cutting down forests also damages the soil. Soil stores carbon, but when trees are removed, the soil becomes dryer and has more exposure to the air. When the carbon in the soil is exposed to the oxygen in the atmosphere, it produces carbon dioxide. (Check out to Chapter 2 for more information about carbon sinks.)

For many developing countries, deforestation accounts for their biggest source of greenhouse gases. These countries clear forests primarily to create land for farming, which causes even bigger problems because agriculture is another major source of greenhouse gases (see the following section for more on agriculture's role in greenhouse gas emissions, and check out Chapter 12 for more about developing nations).

When a rainforest goes under the knife, it has a greater impact on climate change than a northern forest. For example, farmers are clearing large swathes of the Amazon rainforest in Brazil daily to make space for growing soybeans. As we talk about in Chapter 2, tropical rainforests are the most effective type of forest at sucking in carbon dioxide. When they get cut down, Earth loses a climate regulator. The picture gets bleaker still because these forests are often cleared by burning, which puts extra greenhouse gas emissions into the atmosphere. So, when you think globally, you have to think about deforestation as a serious climate problem. Saving the Amazon isn't just about protecting its amazing array of species. It's about saving *humanity.*

For information about how industry can improve land management, check out Chapter 14.

Down on the farm: Agriculture and livestock

We all depend on agriculture to bring us our daily bread — but agriculture involves more than fields of wheat and crops of corn. Food crops are a big part of the agriculture equation, but agriculture also includes growing crops for fiber, such as cotton, and keeping animals for food and other products, such as wool. Of the land that humans use on the planet, 40 to 50 percent of it is used for agricultural practices, the IPCC reports.

Almost every step in agricultural processes produces greenhouse gases — from preparing the soil, to planting the crops, to harvesting the crops, to distributing the food. About 14 percent of world greenhouse gas emissions come from farming practices, according to the World Resources Institute.

Most farm-related carbon dioxide emissions come from how the soil is treated. When it's tilled over and over, and exposed to the air, the carbon stored in the soil is exposed to oxygen in the air and forms carbon dioxide.

When calculating greenhouse gases, scientists don't count agriculture's carbon dioxide emissions. This carbon dioxide is considered *neutral* — the plants that farmers grow suck in just as much carbon dioxide as the agricultural process releases. But scientists do count how much methane and nitrous oxide agriculture produces.

The IPCC reports that agriculture accounts for half of the world's methane emissions and 60 percent of nitrous oxide emissions. Rice paddies alone account for a third of all methane emissions. Livestock, particularly cattle, represent another major source of methane. People in industrialized countries still eat four times as much beef as people in developing nations, but meat consumption in poorer nations is on the rise, which means more animals are being raised, boosting emissions. Regular field crops, such as corn and wheat, produce nitrous oxide emissions, thanks to the large amounts of fertilizer that farmers use to grow them. We talk about how agriculture can produce fewer emissions in Chapter 14, and look at how our individual food choices can help reduce agricultural emissions in Chapter 18.

Chapter 6

Taking It Personally: Individual Sources of Emissions

In This Chapter

- Seeing how travel decisions boost greenhouse gas emissions
- Contributing to global warming by using energy at home
- Heating up the climate with our food
- Connecting waste and global warming

Conversations about the cause of global warming typically focus on the big offenders — the worst industries, dirtiest factories, and scofflaw nations. There's nothing wrong with that. But everyone plays a role in climate change. Each of us uses energy — specifically, fossil fuels — on a daily basis.

From the moment the alarm sounds on the clock radio in the morning until you shut off the TV at night, you're connected to an electrical grid, often fueled by coal or oil. Everyone needs energy to move from here to there — a steady supply of gas for your car or diesel fuel for the bus you ride to work. Most of humanity's food travels great distances before it arrives in our homes, a journey it undertakes thanks to greenhouse gas-producing fuels. (No wonder U.S. President George W. Bush said the U.S. is addicted to oil.)

Don't allow yourself to feel depressed when you realize that your day-to-day choices have been part of the global warming problem. The choices you made were heavily influenced by false pricing and a lack of support for the right choices. Still, the news that all our actions are part of the problem means that all our actions, making changes every day, can be a big part of the solution. If you want to read more on how you can help stop global warming, check out Chapters 17 and 18.

Driving Up Emissions: Transportation and Greenhouse Gases

When Henry Ford innovated the mass production line, he made ownership of the Model T a possibility for average wage earners. The Model T led to faster and bigger cars. And the car transformed the culture of North America. Urban landscapes shifted from focusing on impressive architecture, street cars, and pedestrian access to giving planning preference to cars. Pedestrians and cyclists weren't as important as parking lots and highways. Drive-throughs replaced the corner soda shop. Los Angeles, arguably the apex for car culture, even pioneered drive-through funeral parlors. North America's love affair with the car changed everything — including the atmosphere. Increasingly, the car is also becoming a "necessity" for large numbers of people in the developing world.

About 14 percent of all greenhouse gas emissions come from moving people and goods, according to the World Resources Institute. In Canada and the U.S., the proportion is higher — closer to one third of emissions come from transportation. Almost all our transportation — about 95 percent, says the Intergovernmental Panel on Climate Change (IPCC) — runs on oil-based fuels such as diesel and gas, which explains why transportation accounts for such a large portion of overall emissions. Cars and diesel trucks are the top two offenders, but ships, airplanes, trains, and buses play a part, too. Figure 6-1 shows the breakdown of how each mode of transportation contributes to greenhouse gas emissions. (Refer to Chapter 5 for more information about the problems related to shipping goods.)

Driving

A car engine runs on a simple *combustion system.* Each time you press on the accelerator, gas flows into the engine cylinder, where the spark plug ignites it. The piston goes down, and the crank shaft goes around and turns the wheels. That mini-explosion also creates exhaust fumes that get pushed out of the tailpipe. Those fumes include a healthy dose (figuratively speaking) of greenhouse gases, including carbon dioxide, nitrogen oxides, and hydrocarbons. The exhaust also contains carcinogens (such as benzene), *volatile organic compounds* (solvents that immediately evaporate into the air), carbon monoxide, and other nasties.

Whether you need to run errands or drive the kids to soccer practice, cars, minivans, and SUVs are useful and often perceived as necessary. Most households in industrialized countries own at least one car because most cities and housing developments are built around road infrastructure — making it difficult to survive without one. The majority of people in the developing world still don't have access to a personal vehicle — but that's quickly changing.

China is soaring ahead in private car ownership, which jumped by 33.5 percent from 2005 to 2006, bringing the total to 11.49 million cars, according to the National Bureau of Statistics of China. Nevertheless, with one billion people, China's car population is far smaller than that of the United States.

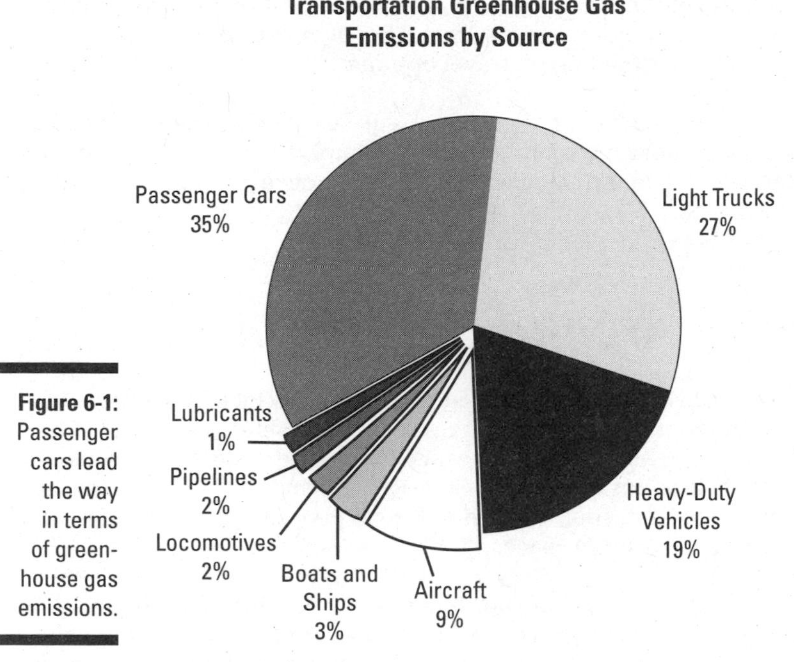

Figure 6-1: Passenger cars lead the way in terms of greenhouse gas emissions.

Flying

Planes burn fuel similar to kerosene, which gives off more emissions than the gasoline in your family car. The IPCC estimates that commercial airline emissions account for 2 percent of worldwide CO_2 emissions — not too bad, except that it's expected to rise to 10 to 17 percent of total greenhouse gas emissions by 2050 at the rate humanity is globe-trotting. People made about 4 billion individual plane trips in 2007, for business or pleasure, and civilization is due to hit 5 billion trips a year in 2010 if it doesn't change its ways.

With countries rapidly developing around the world, international arrivals increased from 800 million to 900 million in just two years, according to the United Nations World Tourism Organization. The people of China are just beginning to tour their own country. They'll begin to travel and work abroad in great numbers soon. Experts predict that China will have 100 million outbound tourists annually by the year 2020, up from 12 million in 2001 and 20 million in 2003. This 100 million would be fourth in the world, behind only the United States, Japan, and Germany.

Flights are becoming cheaper (some European airlines are even offering $1 flights, for which the customer pays only the tax), and when they become cheaper, more of us are able to fly. Today, young adults commonly have flown internationally by the time they're 25. Zoë flew often before she realized the impact of her actions — amounting to a total of over 23 metric tons of greenhouse gases by the time she was 20. Good thing she *carbon offsets* (taxes herself by putting money into a project that will reduce emissions somewhere else) and now takes the train whenever she can. (Check out Chapter 17 for more about green travel options.)

Not only does flying emit a lot of greenhouse gases, it emits them in the atmosphere in a more damaging way. The warming impact of the exhaust from air travel is far worse than the same volume of greenhouse gases emitted on the ground.

Using Energy Around the House

A man's home is his castle. An outdated saying, but the idea that a home is a castle is getting truer and truer. In Canada and the United States, the floor space of the average home is growing while family size is shrinking. House size has real implications for the climate crisis. The bigger the home, the more energy required to heat, cool, and light it. Fewer people are occupying — and heating and cooling — more space.

When it comes to energy use in your home, you can think about it in two ways. First, your *direct energy* use comes from gas or fuel oil that you consume directly — such as the oil-fed heaters or propane gas stoves you may have in your home. And then electricity is sometimes called *indirect energy* because some other energy — oil, hydroelectric, wind, or nuclear power — is used to produce it. How your electricity is produced affects your individual greenhouse gas emissions.

The energy that people use in their homes accounts for 10 percent of greenhouse gas emissions around the world, according to the World Resources Institute. Most of the energy you use directly goes toward heating your home. You use most of your electricity to power your lighting and appliances.

See Figure 6-2 for a complete breakdown of the percentages of greenhouse gas emissions produced from heating, lighting, and other energy uses.

Your energy use may be very different than the average home. For example, you may not have an air conditioner. In Chapter 18, we look at ways to find out where you're using the most energy and electricity, and what you can do to lower your energy use and reduce your dependence on fossil fuels (and save some cash).

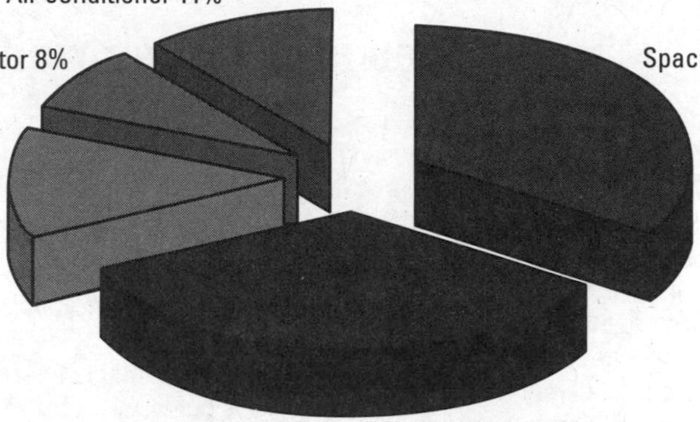

Figure 6-2: Where you use energy in your home.

Climate control

Homes in the U.S. create 150 million metric tons of carbon dioxide every year from heating and cooling alone for 300 million people, according to the U.S. Department of Energy. That's as much energy as Argentina, with 40 million people, uses in one year for absolutely everything.

Heating

Heating takes either direct energy or electricity, depending on whether you have an oil or gas furnace or electrical baseboard heaters. Other types of home heating, such as wood stoves or gas fireplaces, also create emissions. The IPCC notes that burning wood for home heating (and in some countries, cooking, as well) accounts for over 10 percent of energy use in the world. Burning wood adds to greenhouse gas emissions both through the carbon dioxide released during burning and through deforestation.

Older furnaces emit more greenhouse gases than newer models. These old clunkers guzzle fossil fuels, but unfortunately, many homeowners cling to them, worried about the expense of buying a new unit. In reality, these homeowners can save money if they buy a new energy-efficient furnace, which would save them significantly on energy costs — and be less costly to the planet, too.

Cooling

Electricity used to be used only for keeping the lights on. Now, it's what keeps us cool all summer long. In fact, the largest share of home electricity use now goes directly to air conditioning. And in places such as south central Canada, the greater share of power demand has recently shifted from winter to summer. With more 86 degrees F (30 degrees C) days every summer — thanks to global warming — the demand for air conditioning goes up annually.

How green is your electricity?

When it comes to electricity-related emissions, how much you produce depends very much on where you live. If your electricity comes from coal- or oil-fired plants, your electricity use makes a substantial addition to the atmosphere's greenhouse gases. You can find coal plants all over the world, but they're mostly in China, India, and the United States. Some regions, however, don't create very much carbon dioxide because they generate electricity by using energy sources such as hydroelectric, wind, and nuclear power. Canada and Europe have most of the large nuclear and hydroelectric plants. Germany is far ahead in wind power, making up almost a quarter of all installed wind power capacity in the world. The United States, Spain, Denmark, and the Netherlands are also leaders in wind-generated electricity.

Wherever you are in the world, you can go and search for your power provider at `www.carma.org`. This Web site can also show you how much energy your power provider produces, how clean the energy source is, and how many tons of carbon dioxide it gives off.

Chapter 13 covers a lot of the exciting energy alternatives out there.

Only industrialized countries used air conditioning, for the most part, until now. Recent news reports show that sales of home air conditioners have tripled in the last ten years in China. While countries such as China and India move to catch up to industrialized countries, residents are starting to widely use luxuries such as air conditioning. Add warming temperatures into the mix, and you can see a growing air-conditioning trend and a growing demand for electricity to meet that desire.

Traditionally, most Europeans never considered air conditioning. But because killer heat waves have ravaged Europe in recent years, this perspective is changing. For example, the U.K. had to consider new labor laws. In the past, England had laws to ensure a legal minimum temperature so workers could stay warm enough. Because of intense summer heat, they've also had to consider legal maximum temperatures!

Perhaps the most surprising area to need air conditioning is in Canada's far north. Buildings in the Northwest Territories and the Yukon are now being built with air conditioning. The average high temperature in the summer in those territories ranges from 70 to 80 degrees F (in the 20s C), but has been warming up recently and has reached the 90s F (about 30 degrees C).

Electric appliances

Think of every gadget in your home that needs to be plugged in. Actually, it might be easier to think of the things that you don't have to plug in.

Every time-saving device and appliance makes you more reliant on the energy grid.

The refrigerator is the biggest electricity hog in your house — the standard fridge, purchased within the last ten years, can use 120 to 170 kilowatt hours (KWH) a month. (By contrast, a new energy-efficient fridge can use less than 400 KWH per *year*.)

Smaller appliances add up, too — who'd have thought that you can link toasting your waffles to climate change? Your vacuum, microwave, hair dryer, electric kettle, and coffee machine — even your toaster — all matter. Maybe they don't require all that much energy, but you use them almost every day.

Even when you turn your appliances off, many of them are still on! The automatic instant-on features on garage door openers, televisions, and so many gadgets pull power while they wait for you to need them. Manufacturers could make these appliances so that they demand only a fraction of the energy — and California has enacted regulations to insist they draw less power. The California Energy Commission discovered these little instant-on devices used up to as much as 10 percent of home electricity demand! If you don't live in California, your best bet is to unplug televisions, computers, and other electronics when you're not using them. (Or have them all plugged into one power bar that you can turn off with one switch.)

We Are What We Eat: Food and Carbon

Like a warm home in freezing weather, food is a necessity, not a luxury. But sadly, when people sought to make food more accessible and more convenient, and to offer a greater variety, they often did so without considering the environmental toll their innovations might have.

Much of the food that people buy at the grocery store uses a lot of energy to get there — and creates a lot of greenhouse gas emissions as a result. Here are some of the key offenders:

- **Frozen food:** Whether you're talking refrigerated or frozen, these foods burn energy when they're made, while they're being transported, and even when they're sitting in a freezer or cooler in the grocery store (or in your home). The most-energy-used-per-serving prize goes to freeze-dried coffee.
- **Processed and packaged food:** Moving these foods through the production line takes energy, as does making the packaging. (Not to mention the emissions that come from all that packaging when it ends up in a landfill.)

- **Food from afar:** Elizabeth never even saw a kiwi until she was about 18 years old. Her daughter started asking for them for her school lunch in first grade. You may enjoy strawberries and mangoes in the dead of winter, when you can't pick fresh fruit right in your backyard, but moving exotic fruits and veggies around the world by plane, ship, and truck has a real cost in energy. Could people afford them if companies factored in the cost to the climate? And why should your apple be more well-traveled than you?
- **Meat products:** Feeding livestock takes an average of 10 pounds of grain — grain that plays a large role in agricultural emissions — to produce 1 pound of meat. (That fact alone made Zoë adopt a largely vegetarian lifestyle.) Also, when people eat more meat, more land is needed to raise livestock, which often means clearing forests and losing trees that breathe in our carbon dioxide. (Chapter 5 takes a look at farming, and Chapter 18 goes over food solutions.)

Wasting Away

The industrialized world isn't really a "waste not, want not" culture. It's more "Shop till you drop!" And people do shop, creating huge amounts of waste in the process.

Since World War II, household waste from the average U.S. and Canadian home has increased greatly. Over the past 45 years, people in the United States have gone from producing 2.7 pounds (1.2 kg) of waste a day to 4.6 pounds (2.1 kg) a day, according to the Environmental Protection Agency. Key Note Market Assessments (a primary research group providing strategic analysis reports) predicts that worldwide municipal waste production will grow 37 percent between 2007 and 2011.

Changes in marketing and merchandising have significantly added to packaging. Elizabeth is old enough to remember when you went to a local hardware store if you needed nails for a home building project. The person behind the counter would put the nails in a paper bag and weigh them. Zoë's nail-buying experiences have been in modern, barn-like, hardware emporiums. To reduce staff and make shoplifting harder, such stores invented the dreaded bubble-pack approach. To buy a few nails, you buy a big piece of cardboard encased in hard plastic that defies any opening technique except a major attack with scissors. Multiply that packaging approach by a zillion, and you can see how our society wastes so much energy in packaging and why the garbage is piling up.

About 5 percent of worldwide emissions come from waste, according to the IPCC. Landfills rot, adding methane and carbon dioxide to the air. Sewage water is another source of methane — and it adds a dose of nitrous oxide to the mix. Anything that you throw in the trash ratchets up the amount of methane in the landfill, even items such as broken furniture, old toys, and shoes. In some provinces in Canada, coffee cups make up more than a quarter of all materials in the landfill.

Developing countries consume far less per capita, but that doesn't mean they don't have a garbage problem. In fact, it's a different kind of problem — many countries have garbage but no funds for garbage collection or landfills, let alone the luxury of recycling plants. While developing countries are industrializing and building their economies, they're helping increase the amount of waste made in the world. Their wallets are growing, but so are their garbage piles — following in the footsteps of developed countries around the world.

The fast, cheap, and high-emission solution is to simply burn the waste. But burning waste builds up greenhouse gas emissions even further, putting us into a dizzying cycle. Because the volume of waste that civilization produces keeps rising, it has to come up with new ways to deal with it. Many great technologies enable humanity to process garbage back into what closely resembles dirt after only a few years. We go over ways to reduce the amount of garbage you produce in Chapter 18.

Part III

Examining the Effects of Global Warming

In This Part . . .

Climate change could have major irreversible effects on the planet if we don't change our course. The world is already beginning to feel its impacts. This part looks at the spectrum of changes that the climate could bring and how those changes could affect climate systems, the weather, plants and animals, and you.

Chapter 7

Not-So-Natural Disasters

In This Chapter

- Considering the watery consequences of global warming
- Projecting the un-perfect storm
- Figuring out effects on forests
- Feeling the heat waves
- Getting feedback from Mother Earth

Major natural disasters have always happened. Storms, hurricanes, floods, and droughts are all part of the planet's natural weather and climate system.

In the future, however, humanity is going to be facing more and more intense versions of these phenomena — and they're going to be anything but natural disasters. Civilization — or more properly, the greenhouse gases (refer to Chapter 2) that civilization pumps into the atmosphere — will bring them on. Earth could be facing more droughts, hurricanes, and forest fires, heavier rainfalls, rising sea levels, and major heat waves. The excess carbon dioxide that people put into the air might even disrupt the carbon cycle and turn the planet's life-support system into a vicious cycle.

Don't panic, though — you don't need to rush out and build the ark just yet. But this chapter does offer you some very good reasons why civilization needs to start lowering its emissions to cool off global warming.

H_2 Oh No: Watery Disasters

Welcome to the blue planet. Water covers more than 70 percent of Earth's surface. And because of global warming, you might be seeing a lot more of it. Or less. It all depends on where you live.

The relationship between global warming and water is complex. Thanks to rising temperatures, some Antarctic ice is melting and raising sea levels. Elsewhere, glacial melt plays a major role in replenishing the freshwater supply to adjacent farm areas in the spring. As glaciers disappear, this water will disappear, too, increasing drought and water scarcity. Rainfall patterns are altering. Drought is becoming more common.

Rising sea levels

When the planet heats up, sea levels rise for two reasons:

- Antarctic and Greenland ice caps melt into the water. (When ice at the Arctic melts, it does not change sea levels but does change the water's *salinity*, or saltiness.)
- Water expands when it warms.

Even if you take melting polar ice caps out of the picture, sea levels would still rise over the next few centuries up to about 1.3 feet (0.4 meters) for every degree Celsius (1.8 degree Fahrenheit) of temperature change.

Because the majority of the world's population lives along coasts, rising sea levels are one of the most pressing potential effects of climate change. The Intergovernmental Panel on Climate Change (IPCC) predicts that sea levels will rise 7.1 inches to 1.9 feet (0.1 to 0.59 meter) by the year 2100 from both expanding water and melting ice — but this estimate could grow if ice begins melting faster than it is now.

Although this predicted increase might not seem like a lot, it's enough to cover large parts of not only the Maldives (an island nation in the Indian Ocean, which is a scant 1 to 2 meters, or 3.28 to 6.56 feet, above sea level), but also the island state of Tuvalu, 4 to 5 meters (13.12 to 16.4 feet) above sea level. Hundreds of inhabited island nations fear disappearing below the water line.

Figure 7-1 shows that sea levels were steady throughout the 1800s, but that our increased carbon emissions took a rapid toll at the dawn of the 20th century. The right side of the figure shows that the pace is expected to only pick up from here.

Global warming is causing the sea levels to rise in two main ways. First, when Earth's temperature goes up, its water warms — and warmer water takes up more space than colder water. Essentially, the climate change is causing the ocean to expand. Melting ice also plays a role in the rise of sea levels. Extra water from melting glaciers and ice sheets is flowing into the oceans. Two

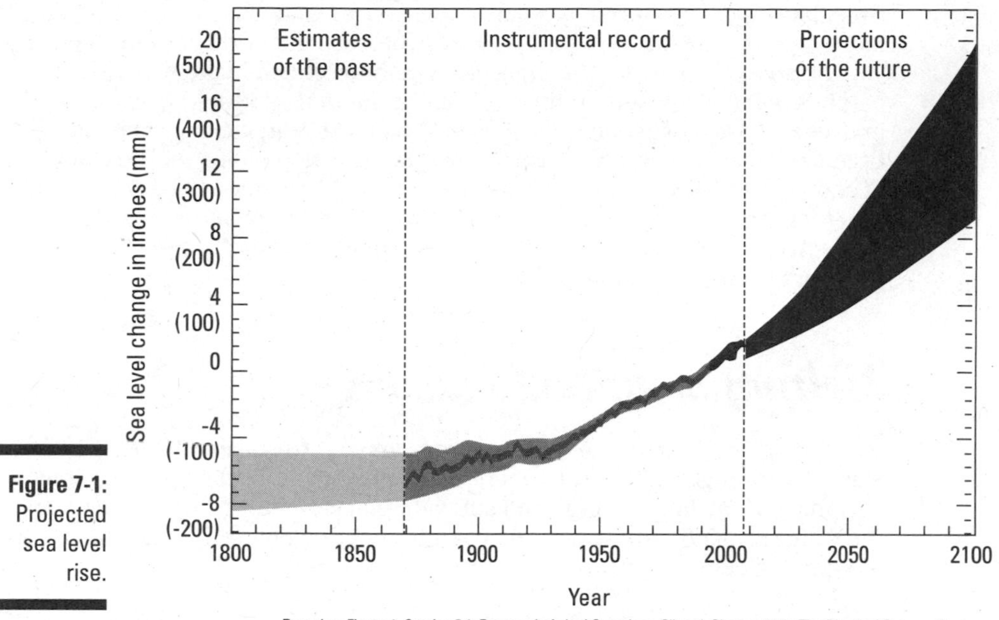

Figure 7-1: Projected sea level rise.

Based on Figure 1, Section 5.1, Frequently Asked Questions. Climate Change 2007: The Physical Science Basis. Fourth Assessment Report. *IPCC. Cambridge University Press.*

areas, in particular, can potentially cause major world changes: the western Antarctic ice sheet and the Greenland ice sheet. If either ice sheet were to collapse, the IPCC estimates that sea level would jump from a predicted 1.9 feet (0.6 meters) by 2100 to 13.1 to 16.4 feet (4 to 5 meters), flooding coastal communities and cities around the world.

Surprisingly, melting sea ice doesn't affect sea levels. The water created by melting sea ice is equal in volume to the ice that was once there. Melting Arctic ice, for example, can impact the strength of ocean currents, such as the Gulf Stream, which could potentially have a serious impact on the climate, but it wouldn't cause sea levels to rise. On the other hand, if enough land ice melts, and that water makes its way into the oceans, sea levels will rise.

Consider the western Antarctic ice sheet — an enormous body of ice that's the size of the state of Texas and contains nearly 10 percent of all the ice in the world. The on-land portion of this ice sheet appears to be weakening because meltwater is working its way underneath and lubricating the base of the glacier, which speeds up its slide towards the ocean. As for the ice over the water, the warming water temperatures directly under it are causing the melting. It's not likely, but it *is* possible, that this ice sheet could collapse. If that happens, the IPCC estimates that sea level rise would jump from a predicted 1.9 feet (0.59 meters) by 2100 to 13.1 to 16.4 feet (4 to 5 meters), putting coastal communities and cities around the world at risk of flooding.

Greenland is also covered by a massive ice sheet, and it's also warming far more rapidly than scientists initially anticipated. The World Meteorological Organization reports that "melting glaciers in Greenland have revealed patches of land exposed for the first time in millions of years." The ice in Greenland is more than 1.86 miles (3 kilometers) thick in some spots. Greenland's ice cover can vary from year to year, depending on the amount of snowfall and other natural weather conditions, but since the 1970s, the ice sheet has experienced a net loss of ice. NASA (the National Aeronautics and Space Administration) reports that the rate of ice loss has doubled in the past ten years in Greenland (see Figure 7-2).

Melting mountain glaciers

Mountain glaciers are large masses of ice carrying rocks and dirt that usually exist at very high altitudes. Glaciers build up from snowfall over very long periods of time. In the spring and summer, glaciers start to slowly melt, and the water runs off into nearby rivers and lakes. See Figure 7-3 for a photo of a mountain glacier.

The IPCC reports that 75 percent of people in the world rely on freshwater from mountain glaciers. In India, for example, a quarter of a billion people depend on a single glacier-fed river.

Greenland Seasonal Ice Melt

1992

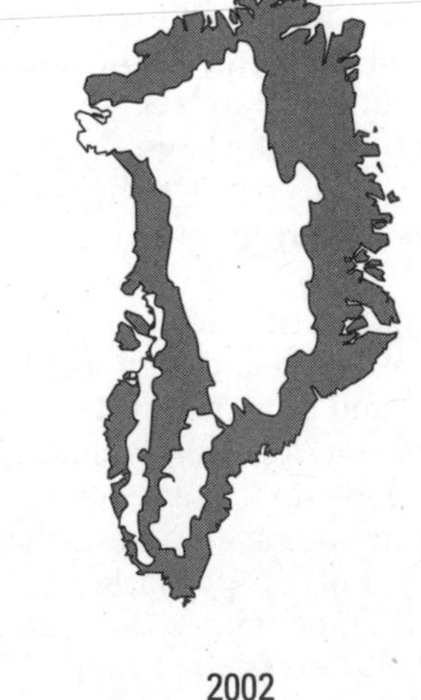

Figure 7-2: Greenland's ice loss over the past ten years.

New opportunities and challenges in the Arctic

Melting northern ice will open up new shipping routes such as the Northwest Passage, which was completely ice-free for the first time in the summer of 2007. Shipping between the Atlantic and Pacific oceans may soon become business as usual. This additional access that the new routes provide has also made oil exploration by ship much easier.

The idea that losing the Arctic ice is "good news" is disturbing given the knock on effects from melting ice. The industry excitement to obtain even more fossil fuels to speed more climate change from under the melting ice suggests a worrying denial of the seriousness of the climate change threat.

But even for those who want more oil and gas, the melting Arctic is not all good news. The Arctic Climate Assessment Council reports that the time available each year for land-based oil exploration has been cut in half because of the warming permafrost (see the section "The Negative Side Effects of Positive Feedback Loops," in this chapter, for more on the impact of melting permafrost). Open water is heating up questions of sovereignty among the countries bordering the Arctic Circle — who controls which waters, and are these newly opened areas the high seas or coastal waters? Additionally, oil exploration brings the risk of spills and other ecological disasters that would harm the fragile Arctic ecosphere.

When this Arctic ice melts, more coastline is exposed, increasing erosion. Erosion also reduces the natural barriers to storm surges. This melting ice is very bad news for indigenous communities that live in the affected areas. We go over the effects that these major Arctic changes have on plants and animals in Chapter 8 and the effects on people in Chapter 9.

Unfortunately, global warming is endangering those water sources. Glaciers are melting more quickly in the spring, releasing a lot of water at once, rather than a smaller, steady flow. This rapid melting can mean floods in the spring,because glacial lakes can't hold all that water at once, and drought in late summer, because the water has drained away. This heavy early runoff will probably also overload the capacity of rivers and streams, eroding their banks and potentially flooding small delta areas with muddy water. Because 300 million people in the world live in deltas, this is a cause for concern (see Chapter 9 for more effects on people).

In addition to accelerating glaciers' melting, higher temperatures are causing most glaciers to retreat. Previously, in colder weather, snow would restore most of the ice that glaciers lost because of melting. But now, because the cold weather doesn't last as long, the glaciers don't get as much snow as they used to. Every year, more ice melts than gets replaced, so the glaciers are shrinking. Glaciers have always advanced and receded, but in the past, they did so veeeeerrry slowly (a few centimeters or an inch a year). Warmer weather is changing that.

Figure 7-3: Mountain glaciers provide freshwater.

Digital Vision

Mountain glaciers around the world are disappearing, from Patagonia to Kilimanjaro, from the Rockies to the European Alps. For example, the 18,000-year-old Chacaltaya glacier in Bolivia has lost 80 percent of its ice area in just 20 years. When Zoë visited nearby the site in 2007, hardly anything was left of it. Glaciers all along the Andes have been the main source of water for cities such as La Paz for centuries. The retreat of most glaciers is a clear physical indication that the world is getting warmer.

Putting a brake on the Gulf Stream

The ocean is always moving. It's not just the tides — a whole system of currents moves water in regular patterns around the world (see Figure 7-4).

Oceans have currents because water varies in density. Cold water is more dense, so it sinks, and warm water is less dense, so it floats. (When you go swimming in a lake, the top is always warmer.) Similarly, saltwater sinks, and freshwater floats. Winds move the top layer of the ocean, which then — simply by friction — moves the layers below it. This combination of temperature, salinity, and wind keeps the currents moving. Meltwater from Greenland, the Arctic, and Antarctica is affecting the ocean currents (check out the section "Rising sea levels," earlier in this chapter, for more information about this melting ice). When this freshwater is released into the oceans, it dilutes the ocean's saltiness. The less-salty water no longer sinks quickly, which can potentially slow the currents.

The Gulf Stream, located in the upper half of the Atlantic Ocean, is part of the overall ocean current system. (If you saw the movie *Finding Nemo,* it's the sea highway that all the turtles and fish cruise along.) The Gulf Stream brings nice warm water from the Gulf of Mexico up past northern Europe. The heat from the ocean warms the air around Europe, which helps explain why Europe tends to be warmer than Canada or Russia, even though it's at the same latitude. After the ocean releases all its warmth on Europe, it continues its route to sweep past chilly Greenland before coming over to Canada (bringing Canada cold air). Figure 7-5 shows the Gulf Stream in action.

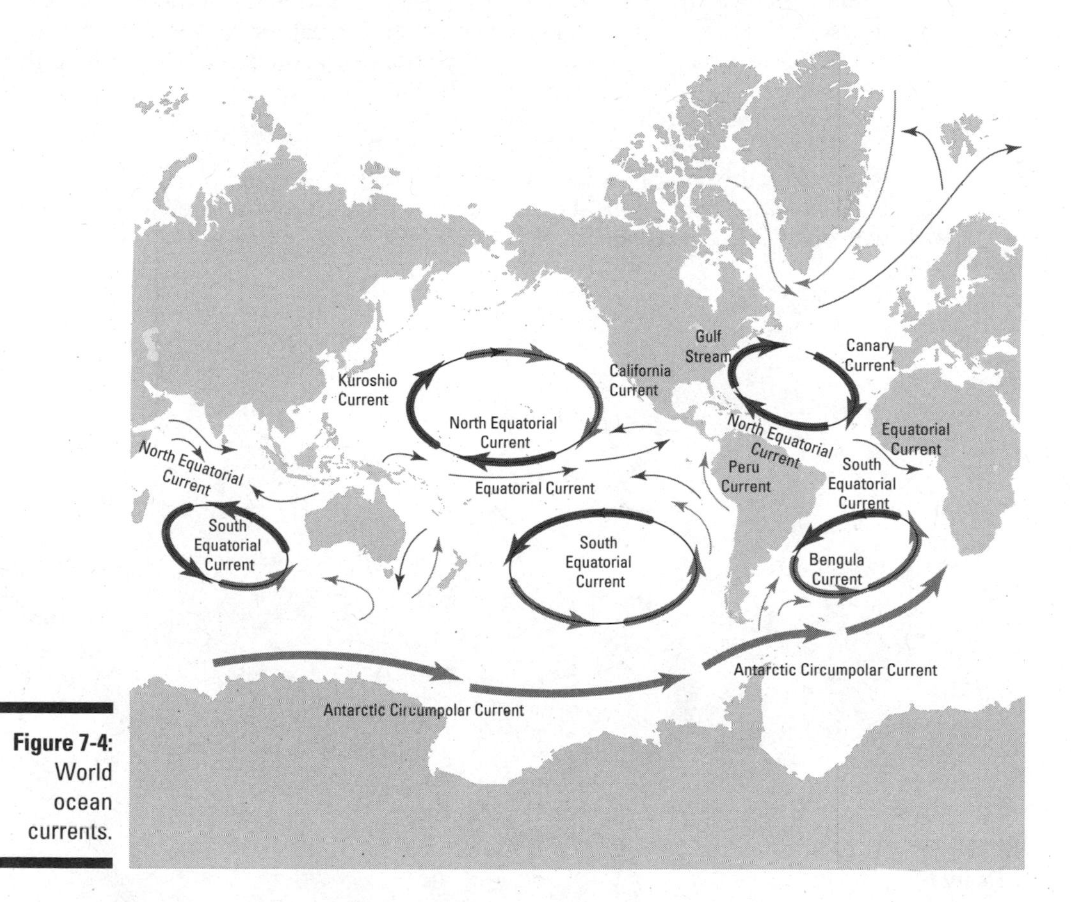

Figure 7-4: World ocean currents.

Some scientists suggest that, over time, the Gulf Stream could slow or even stall because of all the extra freshwater being added to the oceans. The magazine *Nature* recently published research findings from a study done at 25 degrees north in the Atlantic ocean that suggests the Gulf Stream has already slowed down by 30 percent since the last time they checked in the 1950s. The Gulf Stream could stall by 2010.

If the Gulf Stream slows or stops, Europe might start cooling. Sounds tempting because of Earth's rising temperatures — but scientists say that the climate changes triggered by a Gulf Stream disruption would be overall bad news for Europe, unaccustomed to cold winter weather, as well as to the rest of the world. Changing the way the currents work could change how well the ocean sucks up carbon dioxide from the air. (Refer to Chapter 2 for more about the ocean's role as a carbon sink.)

Reports by the IPCC, however, show that the Gulf Stream probably won't stall in this century. But scientists still know very little about how climate change will affect the Gulf Stream.

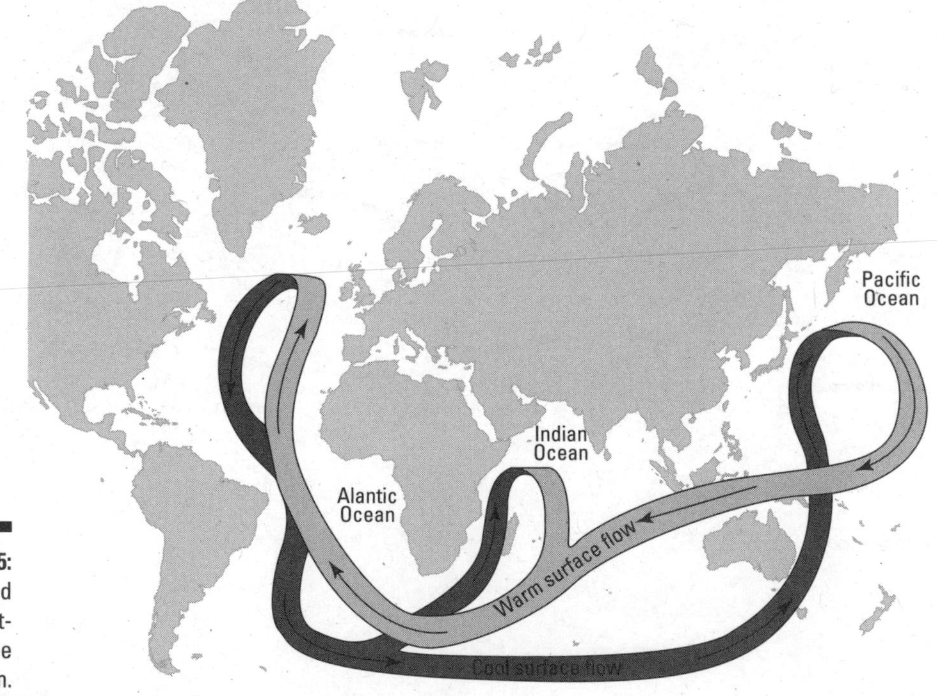

Figure 7-5: Simplified flow patterns of the Gulf Stream.

Rainfall (or lack thereof)

Changes in temperature are altering evaporation and precipitation patterns, which means more rain in some places and less in others. The IPCC says these changes also mean more intense dry spells and rainstorms overall, with high-latitude areas in Europe, Russia, and Canada taking the hardest drenching.

The IPCC reports that inland mid-latitude regions — such as central Canada and inland Europe and Asia — are generally most at risk from more frequent and harsher droughts than what those areas currently experience. Although not in those regions, the land along the Mediterranean in Europe may also experience increased droughts. Droughts and high temperatures put major stress on forests and grasslands; dry, parched vegetation is a fire waiting to happen. The soil suffers, too. Dried out soil can release into the air the carbon that it used to store. (Refer to Chapter 2 for more about how soil contributes to carbon dioxide in the atmosphere.) Drought is hard on people and animals, because all living things depend on water.

Deserts around the world are expanding. The Gobi desert in China, for example, is approaching Beijing — China is already 30 percent desert, expected to soon become 40 percent. Although expanding deserts are a natural phenomenon and not directly linked to global warming, the combination of increasing desert area and droughts can have negative effects of stressing freshwater sources and food production. The IPCC reports that the duo of natural warming and human-caused warming has caused the number of dry areas around the world to double since the 1970s. China is working to combat the expansion of the desert by planting forests.

Flooding

Three climate change consequences lead to flooding: rising sea levels, quicker-melting snow and glaciers, and more intense rain showers. The IPCC expects that the rising sea levels and harsher rainstorms will increase the number of floods in many places, including both *flash floods* (floods that happen very suddenly, often because of heavy rainfall and/or the ground is so dry it can't quickly absorb the rain) and *large-scale floods* (floods that stick around for a while, caused either by prolonged rainfall or water that can't drain away easily).

It is quite typical of climate change scenarios to predict that average annual precipitation will remain nearly constant, but that areas will experience long periods of drought followed by an enormous volume of rain. A good example of this occurred in Mozambique when a persistent drought lasting months immediately preceded the torrential rains of 2000. That nation's annual precipitation fell within days on the dry and desiccated lands.

The most likely areas to experience more flooding are high-latitude countries, such as the United Kingdom.

At the time this book went to press, the IPCC saw increased flooding as a future event. At this point, scientists can't blame existing levels of climate change for any major changes in flooding, but it may only be a few more years before scientists trace the recent floods to climate change.

Freshwater contamination

Most of the creatures that walk (or crawl or slither) on the planet require *freshwater* (water that isn't salty) to survive. Unfortunately, flooding and rising sea levels, two of the effects of climate change, pose two contamination risks to freshwater:

- **Getting it dirty:** Runoff from flooding can get into drinking water. This runoff washes over city streets and can take anything with it — from any dirt or garbage in the streets, to overflowing sewer systems, to pesticides and fertilizers from our lawns.
- **Getting it salty:** The higher sea levels mean saltwater intrusions. Not only can saltwater get into fresh surface water, it can also work its way down into *aquifers* (water-bearing rock, which can provide well water) and coastal freshwater rivers. More people live along coastlines than any other region, and those people have the fewest sources of freshwater. Seawater contamination will only worsen that state of affairs.

Stormy Weather: More Intense Storms and Hurricanes

You may have heard about stronger storms and hurricanes as an effect of global warming, either on the news or from watching Al Gore's documentary *An Inconvenient Truth.*

Global warming is heating up our oceans. In fact, the IPCC reports that oceans have absorbed about 80 percent of the heat from global warming. Hurricanes are now occurring in the top half of the northern hemisphere, such as Canada, because of these warmer ocean temperatures, particularly at the surface. Historically, colder ocean surface temperatures in the north slowed down hurricanes, turning them into powerful, but nowhere near as destructive, tropical storms. Now, however, the water's warmer temperatures don't impede storms. In fact, warming up surface water is like revving the hurricane's engine.

The number of tropical storms and hurricanes hasn't increased. In fact, that number has stayed fairly uniform over the past 40 years, the IPCC reports. The intensity of tropical storms and hurricanes, however, *has* increased. For example, eight Atlantic Category 5 hurricanes have occurred so far this decade; no other decade on record has had so many. Hurricanes hit Canada's two major coastal cities, Halifax on the east coast and Vancouver on the west, in 2003 and 2006, respectively. In fact, 2003's Hurricane Juan was the first full-force tropical hurricane ever to hit Atlantic Canada.

These bigger storms and hurricanes bring rougher coastal storms, bigger storm surges, higher water levels, taller waves, more storm damage, and flooding. Some storm-protection barriers might not be strong enough to protect against the hurricanes that are coming, and some cities might need to reevaluate their protection. (Think New Orleans!)

The most recent science shows that storm and hurricane intensity has grown around the world since 1970. This rising intensity is linked to rising ocean surface temperatures. But some scientists have challenged these data because they're not in line with climate models; in fact, some climate models predict that storms and hurricanes are about to become less intense. Despite this disagreement, people are better to be safe than sorry when so much is on the line. Protecting humanity means reducing greenhouse gas emission immediately as well as better preparing for storms by building better protections and improving our response to natural disaster emergencies.

Forest Fires: If a Tree Dries Out in the Forest

As we discuss in Chapter 2, forests are critical in keeping excess carbon dioxide out of our atmosphere. Unfortunately, the number of forest fires is greatly increasing, and global warming is the cause.

The increase in hot, dry weather means dryer forests, ideal fodder for fire. Forest fires around the world last longer and burn with more intensity than previously recorded. The area of land burned by wildfires has surged in the past 30 years across North America. The IPCC reports that a one-degree Celsius rise (1.8 degrees Fahrenheit) in average temperatures can increase the length of the fire season in northern Asia by 30 percent.

These fires have serious consequences, not only for the environment, but also for infrastructure. Major wildfires in Canberra, Australia, in 2003 ruined 500 houses and cost a hefty 261 million dollars (U.S.) in damage alone. The 2007 fires across the state of California destroyed 1,500 homes. The IPCC expects forest fires to increase while temperatures continue to rise and some areas experience reduced rainfall.

Warmer weather also means more pine beetles in the western U.S. and Canada. The pine beetle is an insect that has a special talent for turning a forest into firewood. Previously, pine beetles didn't survive the winter. Now, their numbers grow annually. (See Chapter 8 to get better acquainted with pine beetles.)

The increasing numbers of pine beetles have significantly affected interior British Columbia, where Zoë grew up. The Montane forest, where lodgepole pine predominates, has lost an area of forest the size of two Swedens because of climate-induced beetle attacks.

In fact, because of both fire and increased insect damage, forests in Canada ceased to be a net sink (refer to Chapter 2) for carbon in the mid-1970s. Canada's forests still hold millions of tons of carbon, but on an annual basis, these forests now give off more carbon than they suck in. Adding to this vicious cycle, forest fires pump carbon dioxide into the air when the wood burns and releases the gas.

Turning Up the Heat

You may think that we're just stating the obvious when we say that global warming will bring about more hot days and warm nights. But those hot days can be fatal, particularly when they constitute a *heat wave,* a prolonged period of very hot weather. In fact, in both the U.S. and Europe, heat waves kill more people each year than tornadoes, floods, and hurricanes combined. An estimated 35,000 people died in Europe because of extreme heat waves in the summer of 2003. During a five-day heat wave in Chicago in July 1995, several hundred died. You don't necessarily hear about other heat waves: Many countries across Africa have been enduring heat waves that last for longer periods of time than they have in the past.

High temperatures can mean high stress on the body, particularly heatstroke. Heatstroke, in extreme cases, can lead to chronic illness and sometimes death (see Chapter 9 for more on global warming's effects on people). Society's most vulnerable — the poor, the elderly, and (especially) the elderly poor are usually the victims of killer heat.

Heat waves also claim livestock. Heat stress can lower livestock's ability to reproduce and increase mortality rates. News reports show that the 2006 heat wave that hit California killed 25,000 cattle and 700,000 chickens and turkeys.

Unfortunately, the future of heat waves is scorching. Heat waves will continue to get more intense and last longer each time they occur. Los Angeles can expect its regular 12 days of heat waves per year to jump to between 44 and 95 days of heat waves a year by 2070 to 2099.

And when people want to cool off, they're adding to another problem — air conditioning and refrigerators working overtime commonly raise the amount of electricity used in cities during heat waves. Climate change will raise peak energy demands because people will reach for the air-conditioning dial, making conservation efforts more difficult.

The Negative Side Effects of Positive Feedback Loops

The carbon cycle, which we talk about in Chapter 2, is like our planet's respiratory system. Some organisms emit carbon, and others breathe it in. When the carbon cycle is balanced (as we know it), it's a natural wonder, ensuring that the air doesn't contain an excess of carbon dioxide.

When the carbon cycle goes off-kilter because of too many carbon emitters and too few carbon absorbers, it could prove disastrous. Climatologists fear that an imbalanced carbon cycle will create positive feedback loops. A positive feedback loop is anything but good news, even if it does have the word "positive" in its name. In a *positive feedback loop,* effects are perpetually amplified. In the case of the carbon cycle, constantly escalating carbon emissions cause ever-increasing temperatures. If the feedback loops accelerate, there is a theoretical possibility of a *run-away greenhouse effect.*

A dim hope to fight global warming

When a great deal of *particulate matter* — such as volcanic ash, dust, pollution, and gas from aerosols — prevents some of the sun's light from reaching Earth, it's referred to as *global dimming.* For a while, scientists hoped that global dimming might help correct global warming — less sunlight reaching the Earth would cool things down considerably. But NASA reported in 2006 that the amount of particulate matter in the atmosphere is decreasing — which could be due to the effect of stricter air quality regulations on industry — so more of the sun's light reaches the planet.

Here's an example of the run-away greenhouse effect in action: A high amount of human-produced carbon dioxide in the atmosphere intensifies the greenhouse effect, leading to higher temperatures that melt the once permanently frozen ground (known as *permafrost*) in the Arctic. That frozen ground has been a greenhouse gas reservoir for thousands of years, storing methane that it releases into the atmosphere when it melts. And that additional methane heats things up even more, which causes more carbon emissions. (You can see how this situation gets into a worsening loop.)

The same kind of vicious cycle happens when Arctic ice melts. When the ice is present, the sun strikes the white surface and bounces back, just like a mirror reflecting light. You know the handy trick for dressing on hot days? White clothing keeps you cooler because white reflects the sun's rays, and black clothing soaks in the heat. Well, Arctic scientists call this the *albedo effect.*

When the ice melts, dark ocean water replaces it. The dark ocean water doesn't reflect the sunlight. (Just like your black t-shirt doesn't keep you cool.) It soaks in the heat, warming ocean water even faster and melting the ice more quickly, which leads to more open water and more warming. Warmer waters might also mean that other organisms can thrive where they haven't before, pushing out algae and lowering the amount of carbon dioxide being sucked up. Figure 7-6 shows the sizeable contrast between the white ice and the comparably black water — you can imagine how much more heat the water absorbs than the ice does.

Positive feedback loops also apply when forests catch fire because of warm and dry conditions. Those burning forests release the stored carbon in their branches, trunks, and leaves, adding to the carbon dioxide in the atmosphere. That additional carbon dioxide leads to more warming and dryer conditions, which lead to more forest fires.

Not so permanent

Unfortunately, some permafrost is already melting in the western Arctic of Canada and Russia's Siberia. When permafrost melts, it releases into the atmosphere methane gas that had been stored for thousands of years. The melting alone is bad news for local people. Roads collapse when the ground subsides. Houses and even villages have had to be relocated.

The *Inuit* — indigenous people of the circumpolar region, known largely as Inupiat Eskimo in Alaska and Inuit in Canada and Greenland — used to put foods that needed refrigeration into the permafrost, but they can no longer do that because the ground is warming up and the permafrost is disappearing. Their natural fridge is defrosting.

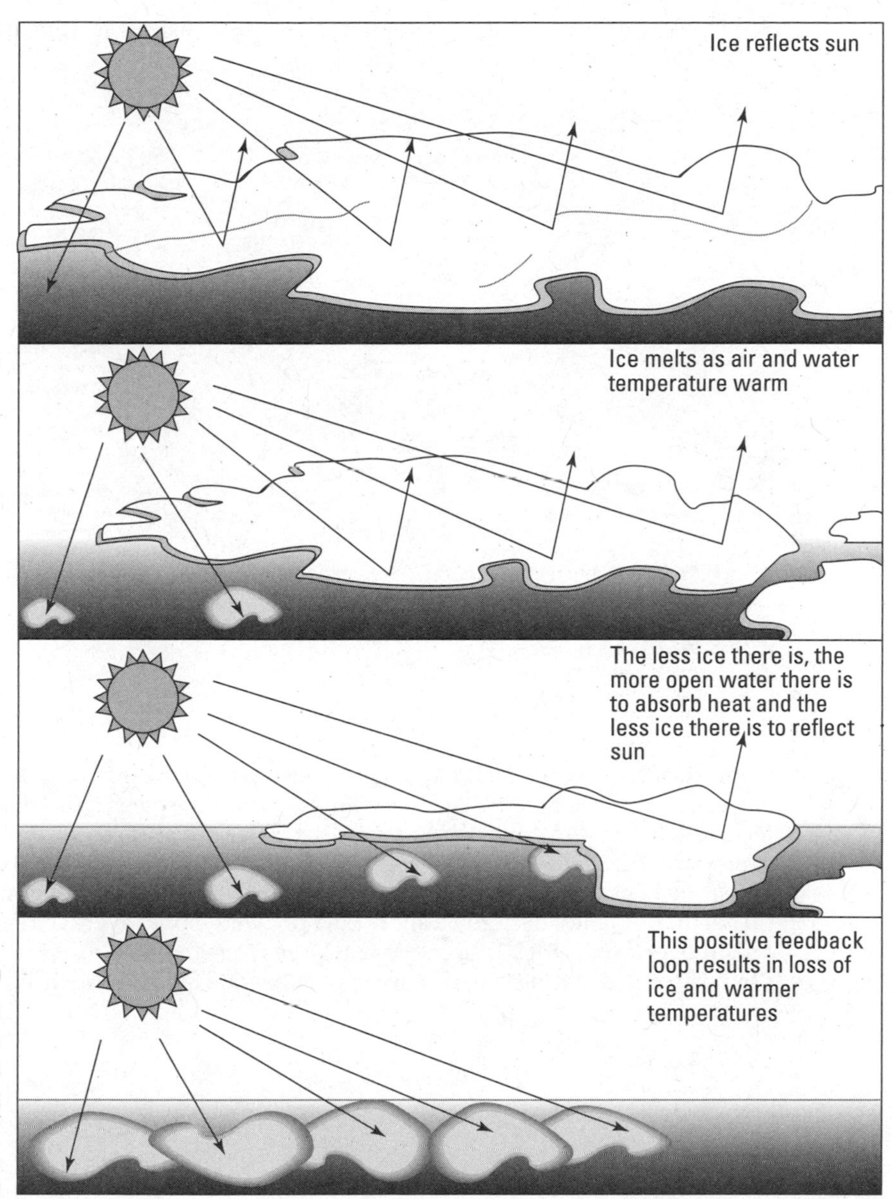

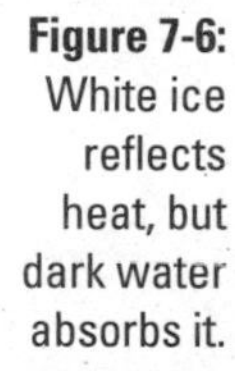

Figure 7-6: White ice reflects heat, but dark water absorbs it.

See Figure 7-7 for a visual representation of other positive feedback loops.

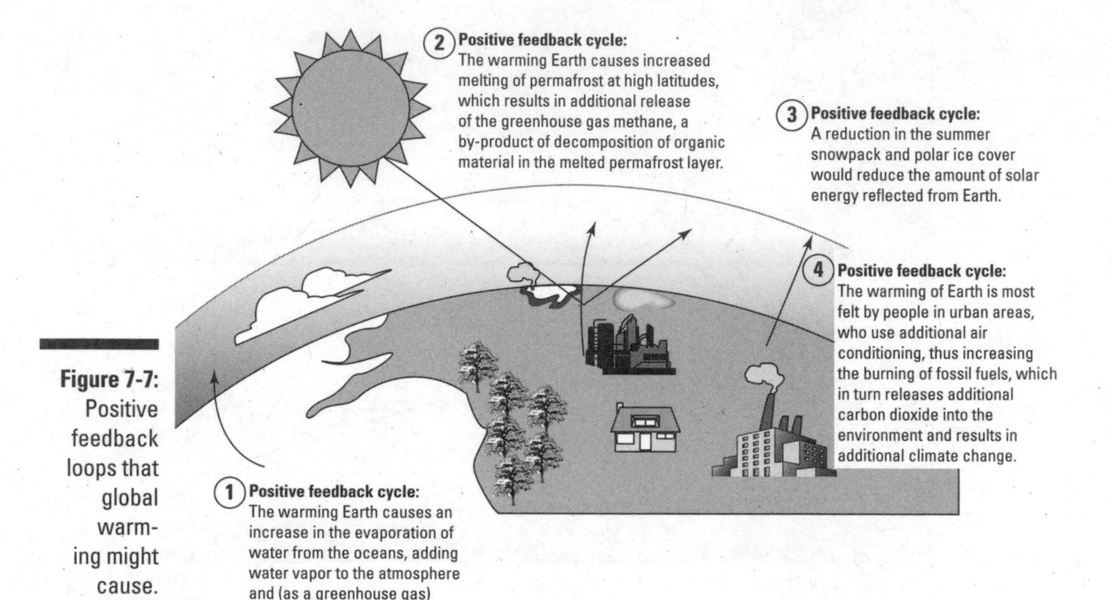

Figure 7-7: Positive feedback loops that global warming might cause.

SITUATION CRITICAL

Some scientists worry that if humanity doesn't dramatically reduce greenhouse gas emissions, the positive feedback loops could take over. If the positive feedback loops take control, reservoirs of carbon that represent thousands of years' worth of carbon sucked from the atmosphere would rapidly release their greenhouse gases. A run-away greenhouse effect could, theoretically, end life on Earth. Humanity has only a short time to avoid global average temperature rise to over 3.6 degrees Fahrenheit (2 degrees Celsius) above 1850 at which point the risk of a run-away climate impact becomes dangerously likely.

Chapter 8

Risking Flora and Fauna: Impacts on Plants and Animals

In This Chapter

- Understanding the possible pressures on the environment
- Watching for a sea change
- Observing the effects of climate change on the world's forests
- Taking a look at how climate change will affect animals and plants

Some people call it nature. Some talk about the birds and the bees, all creatures great and small. Others talk about flora and fauna. Scientists call the living things on Earth, from the genetic level to the landscape level, biodiversity.

Biodiversity is also called "biological diversity" — the planet's variety of living species. The world is abundant with diversity, from the deep forests of China, to the mountains of Canada, to the icy waters of Antarctica.

Biodiversity is affected by a deep matrix of issues — ranging from the impact of people, to the way species interact with each other, to changes in the environment around communities. Climate change has become another issue added to the mix.

The United Nations Convention for the Protection of Biological Diversity has identified climate change as a threat to biodiversity. While the natural world becomes increasingly threatened, the way humans use and interact with plants and animals will change. More or less, all the living things on the planet are in this together. In this chapter, we look at what kind of pressure climate change is putting on the Earth's biodiversity, whether organisms can adapt to these changes, and which plants and animals will likely be most affected.

Understanding the Stresses on Ecosystems

If you've ever watched a nature program on television, you know that the natural world isn't the most peaceful of places, with predators mercilessly chasing down prey. But if you've been out on a hike, nature may appear to actually be a pretty harmonious, finely balanced arrangement. The relationships of living creatures with each other, the land, and the climate within a particular area are known as an *ecosystem*.

Ecosystems can be highly adaptable. For hundreds of millions of years, the Earth's ecosystems have adapted to changing climates, responding to differences in rainfall, shifting temperatures, available land, and even changes in the levels of carbon dioxide in the air. But now, climate change is occurring at such a rapid pace that many species likely won't be able to adapt, and entire ecosystems may be transformed.

Worse still, global warming is happening in combination with many other human-created pressures on ecosystems. With 6.5 billion people on the planet, and on the way to 9 billion people by 2050, ecosystems are under much greater pressure than they ever were before. If you've ever looked out the window of a plane while flying over land, you've seen manmade patches and lines across the land — patches for developments such as cities, farming, and logging; and lines for structures such as highways, roads, and railroad tracks. Every patch or line that people make fragments the natural ecosystems. Often, people conserve a patch of forest, but human development surrounds that forest, so the plants and animals must rely on each other in that reduced area, allowing less room for adaptation. Air and water pollution hurt biodiversity, too. Humanity has driven many species to extinction.

Faced with a change in their environment (for example, warmer temperatures), species often adapt by moving to a climate to which they're better suited, but plants and animals that live on small islands, on mountain peaks, or along coastlines often don't have this option. Species living in areas that have been fragmented by human developments — forests surrounded by subdivisions, for example — also can't easily move.

Some species can't move very far, no matter what. Slow and steady won't win the race for survival if snails and turtles have to follow their ideal temperature ranges toward the Earth's poles. Think about flightless birds and insects — they're in for a very long walk on very small legs. And most plants stick to their roots (although some plants do manage to travel — see the sidebar "But plants can't move!" in this chapter).

Some plants and animals can live only in certain temperature ranges. For instance, Alpine meadows are very dependent on cool temperatures along the tops of mountain ranges. Coral reefs can bleach out and die because of a very small temperature change in the surrounding ocean. In Australia, some species of eucalyptus trees can survive no more than a 1.8 degree Fahrenheit (1 degree Celsius) temperature shift.

The Intergovernmental Panel on Climate Change (IPCC) says that an average increase in global warming of 3.6 degrees Fahrenheit (2 degrees Celsius) above 1850 levels — or 2.2 degrees Fahrenheit (1.2 degrees Celsius) above today's temperatures — will have serious effects on all the world's major ecosystems, both on land and in water. (This shift in temperature is sometimes called the danger zone, which we discuss in more detail in Chapter 3.)

Ecosystems typically take from a few years to a few centuries to adapt to climate changes — which means that some ecosystems may not be able to cope in their current form. Even if alligators used to swim in the swamps of the current-day Arctic, never before have changes happened at such rapid rates and in combination with so many other pressures that human society has placed on ecosystems. The UN Convention for the Protection of Biological Diversity specifically states that "climate change is likely to become the dominant direct driver of biodiversity loss by the end of the century."

Scientists don't fully understand how ecosystems will be affected by continuing climate change. Ecologists don't know how organisms will respond to the stresses on their ecosystems. Climate change will affect each and every species in different ways. Some species may need to move, others may need to start eating different foods.

But plants can't move!

We know, we know — we keep talking about plants and animals adapting by moving elsewhere. But how does a plant or tree move? Of course, individual plants can't travel, but the seeds that the plants spread can. Here's how:

- **Poop:** Trees or plants that grow fruit or berries (think apple trees or raspberry bushes) are a food source for animals and people. When you eat a raspberry and swallow the seeds, well . . . what goes in one end must come out the other! Animals in the wild eat the fruit, wander while they digest, then do their business. The seeds in the poop decompose into the ground and grow a raspberry bush.
- **Picked up, dropped down:** Some fruits, such as plums and peaches, have big pits. Birds and animals take the fruit, eat it somewhere else, and discard the pit — which potentially may blossom into a new tree.
- **Wind:** Plants produce pollen and seeds that, after they develop, get carried away by the wind. Sometimes, they don't go far, but they can! For example, a tree drops pine cones, and those pine cones can also be spread by the wind.

Healthy ecosystems can help humanity adapt to climate change and even reduce greenhouse gas emissions. As we discuss in Chapter 2, plants and healthy soils take in carbon dioxide. Plants create a lower temperature, and their root systems keep the soil healthy and able to absorb water, instead of letting it run off. The Convention on Biological Diversity reports that healthy and diverse ecosystems are more likely to adapt to climate change than are those that aren't healthy or diverse.

Warming the World's Waters: Threats to the Underwater World

Water makes up 70 percent of the Earth's surface, making it a very important set of ecosystems, including oceans, seas, wetlands, rivers, streams, and swamps. Climate change will affect all of these ecosystems in the form of increasing water temperatures, rising sea levels, or droughts brought on by rising air temperatures. (Refer to Chapter 7 for more about the natural disasters global warming may cause.) Exactly how these ecosystems will be affected, no one knows. Climate change is reshuffling the deck of water systems, and the world doesn't know what kind of hand it'll get dealt.

Many fish species are already at risk of extinction due to overfishing. According to scientists at Dalhousie University in Canada, fish populations have dropped more than 30 percent since the 1950s and are continuing to decrease. The projected loss from climate change will further elevate the risk of extinction.

Whether a few species are thriving or many species are declining in an ecosystem, these differences change the way the ecosystem functions. Each organism plays a role in an ecosystem. Ecosystems are remarkably adaptable; change the role of one organism, and the whole system alters in response. Global warming–related changes to ecosystems may cause some radical changes in their composition and how they function.

Under the sea

Ocean ecosystems are very complex. Climate change could change these ecosystems dramatically, proving disastrous for some species. The World Conservation Monitoring Centre highlights pressures brought on by global warming that they project will particularly affect ocean life:

- **Increasing carbon dioxide in the water:** When the ocean water takes in carbon dioxide from the atmosphere, the carbon dioxide dissolves, becoming *carbonic acid,* which makes the water more acidic. Raising the water's acidity negatively affects shell building for sea snails or coral, and when those species are endangered, so are all the species that prey on them or (in the case of reefs) live in and depend on them.
- **Increasing water temperature:** Many ocean species are sensitive to the temperature of the water. Coral reefs, for example, have been shown to suffer badly from higher water temperatures. The Great Barrier Reef off the coast of Australia is at risk.
- **Shifting ocean movements:** The ocean is constantly in motion, with a vast conveyer belt of currents that carry warmth to cooler parts of the globe (and vice versa). These currents provide food sources for other sea creatures. Increased fresh meltwater from polar ice, brought on by climate change, has the potential to shift, stall, or stop ocean currents altogether. (Refer to Chapter 7 for more information on the potential impact on ocean currents.) Currents influence the heat transfer in ocean environments, and changes in how warm water moves around can have consequences for temperature-sensitive species.

The impact of these changes will be different throughout the oceans of the world.

Drops in productivity

The IPCC expects icy ocean ecosystems in the Arctic and Antarctic to drop in productivity by 42 percent and 17 percent, respectively, by the year 2050. A productive ecosystem is one that's conducive to life; while an ecosystem's productivity drops, organisms within it decline. Much of this expected drop in productivity in the polar regions of the world relates to the expected decline in phytoplankton. *Phytoplankton* is a microscopic plant that's called algae when it's clumped together (but it isn't the same thing as blue-green algae, which are bacteria), and it's the primary food source for the entire ocean food web.

Phytoplankton in ice ecosystems forms along the edge of the sea ice. When sea ice shrinks, so does its edge, limiting the area that produces phytoplankton. This reduction in phytoplankton will affect the fish and krill that feed on phytoplankton, their predators (penguins), and *their* predators (seals). See Figure 8-1 for a simplified version of the ocean food chain.

Research from the Institute of Science in Society notes that, without phytoplankton, "marine life will literally starve to death."

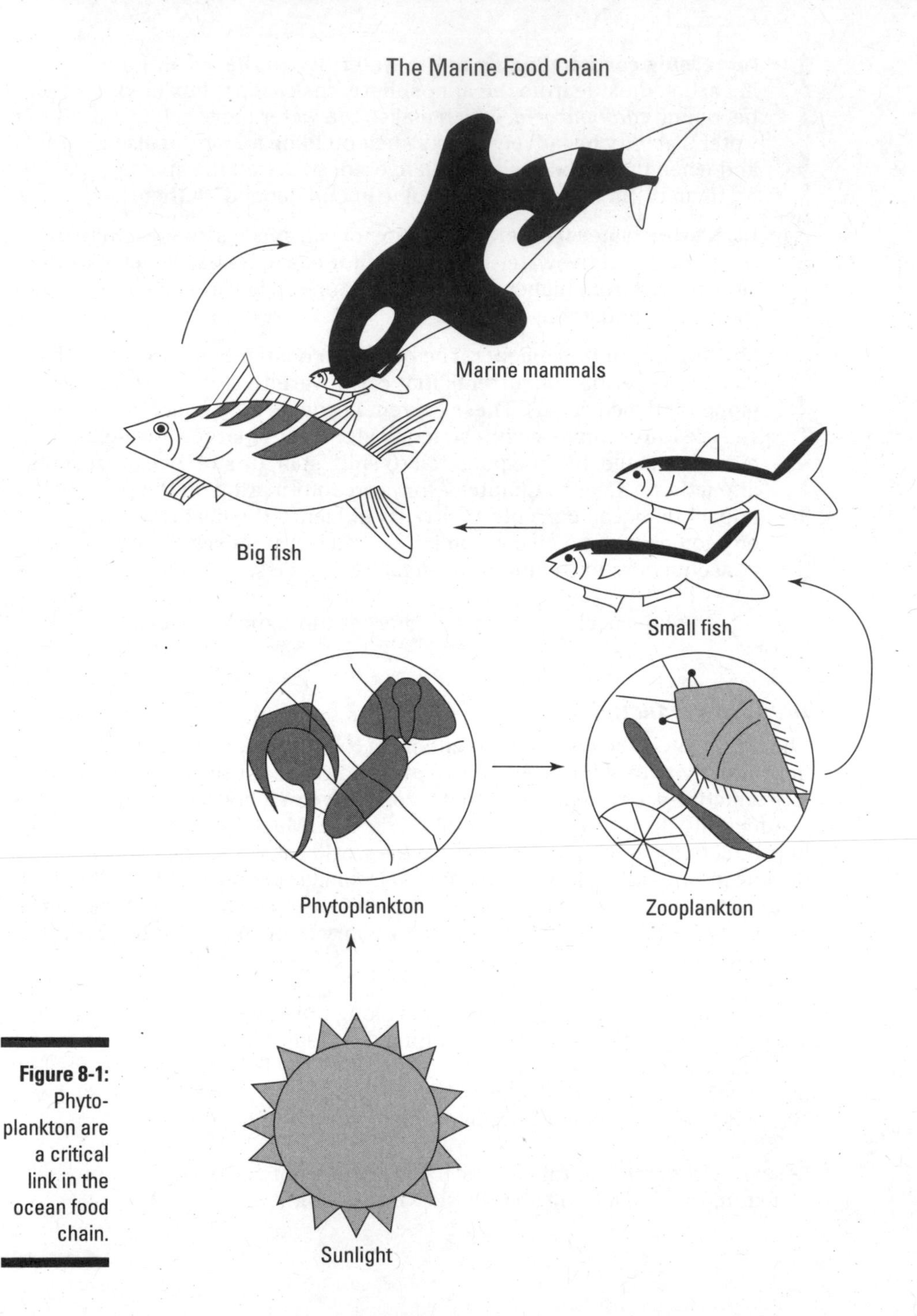

Figure 8-1: Phytoplankton are a critical link in the ocean food chain.

Coral at risk

Coral reefs are already among the marine ecosystems at greatest risk of being destroyed by the effects of climate change. They're the hub of diversity in the oceans. Coral might look like a rock or a rocky plant, but what you're seeing is actually the outer skeleton of a living animal — yes, an animal — that has a stomach, a mouth, and even a sex life. Corals come in all shapes and sizes, twisting from left to right and in a rainbow of colors. They live all over the world, in both warm and cold waters, but they're mostly concentrated around Southeast Asia and the Caribbean. The reefs support a wide array of small living water plants and fish. Without the reef, the plants and fish have no home. And without the plants and fish, the bigger fish in the sea have no food, and so on.

The IPCC reports that most, if not all, coral reefs will be bleached (meaning dead) if global average temperatures rise 3.1 degrees Fahrenheit (1.7 degrees Celsius) above 1850 levels — or 1.6 degrees Fahrenheit (0.9 degrees Celsius) above current levels. A sea surface temperature rise of 3.6 to 5.4 degrees Fahrenheit (2 to 3 degrees Celsius) above 1850 temperatures — or 0.4 to 4 degrees Fahrenheit (0.2 to 2.2 degrees Celsius) above current levels — is the danger zone for coral, unless it can adapt. With a temperature rise of 5.4 degrees Fahrenheit (3 degrees Celsius) above 1850 levels — or 4 degrees Fahrenheit (2.2 degrees Celsius) above today's levels — all coral reef systems would die.

Despite the increasing evidence linking climate change to coral reef destruction, you can't easily separate the effects of the climate from other major human pressures, such as pollution and fishing. The Caribbean, for example, has already lost 80 percent of its coral reefs because of other human pressures.

The IPCC predicts that areas covered by cold-water corals, such as those found off the east coast of Canada and the west coast of Norway, will go through a huge drop in productivity by the end of the century. Cold-water corals depend on nutrients that sink from the surface or arrive via ocean currents, so changes in ocean currents could be a threat to their survival.

Dramatic ecosystem changes

The IPCC expects a major change in the way marine ecosystems function worldwide. While some species decline and others migrate, the animals and plants in the sea will form new relationships. Predators may find themselves prey, other species may wind up competing with one another for the same food resources. No one knows how easy these transitions will be.

Lakes, rivers, wetlands, and bogs

The world's freshwater will also see some serious changes brought on by global warming. Earlier spring runoff from snow and glacier melt, and less runoff in late summer, will affect rivers in many parts of the world. Compared to now, these rivers will be at higher levels in the spring and at lower levels in the late summer. And the lower summer levels mean less freshwater availability when it is most biologically needed. (Refer to Chapter 7 for more about melting glaciers.)

Wetlands and bogs are also at risk. These ecosystems depend on having enough water, and they don't have much room for adaptation. When temperatures get warmer, more water evaporates from these areas. A dry bog or wetland is a bog or wetland no more.

Declining water quality

When water temperatures in lakes and rivers go up, the water quality declines. Here's the process that leads to reduced water quality:

1. Warm waters are good homes for *algae* — tiny, green, plant-like organisms that look like scum on water.
2. When the algae fall to the bottom of the waterbed and decompose, underwater sediment gives out phosphorus. *Phosphorous* is a poisonous chemical that looks like yellow wax and glows in the dark in natural environments. It's also used for man-made products, such as the red bit on the tips of matches. Phosphorous can give a bad taste and odor to drinking water.
3. Algae thrive on phosphorous. Excess phosphorous means more algae.
4. Algae grow quickly, called an *algal bloom.* Bacteria and fungi decompose the algae and hog all the oxygen in the water because they breathe it in. The oxygen level in the water falls, meaning the water contains less oxygen for fish — which need oxygen to survive, just like we do.

Decreasing fish species

Scientists expect that warmer temperatures in lakes and rivers will have an adverse effect on fish species. Although some species may thrive — indeed, some fish species populations are actually increasing — others potentially face extinction. Different species have different ranges of temperatures in which they're comfortable. Species that have large comfortable temperature ranges can adapt more easily than more sensitive species because they're not affected by these temperature shifts.

As the saying goes, you win some, you lose some. Unfortunately, climate change means the Earth will lose a lot. For example, the IPCC predicts that if global temperatures rise 2.3 to 3.1 degrees Fahrenheit (1.3 to 1.7 degrees Celsius) above 1850 levels — or 0.9 to 2.2 degrees Fahrenheit (0.5 to 0.9 degrees Celsius) above current levels — then North America will lose 8 to 16 percent of its freshwater fish habitat. This loss of habitat translates to a 9- to 18-percent loss of salmon.

Risking Our Forests

At least at first, warming temperatures are expected to be good for plants and trees. Already, the IPCC reports, forests have been more productive over the last few years than in previous decades. The growing seasons are becoming longer, and the air contains more carbon dioxide for the plants and trees to take in.

In time, however, rises in temperatures and changes in ecosystems will have an overall negative effect on forests because of larger, more frequent fires; regular disease outbreaks; and insect infestations.

Tropical

Rainforests, which are most commonly found in the tropics, can survive only within small temperature ranges. As their name suggests, rainforests depend on rain. The combination of warming temperatures and changing rainfall patterns brought on by global warming could adversely affect the rainforests. In that situation, the IPCC predicts that species in tropical mountain forests face a high risk of extinction. IPCC reports show that the Amazon region could lose a lot of its forest and biodiversity if global average temperatures reach 4.5 degrees Fahrenheit (2.5 degrees Celsius) above 1850 levels — or 3.1 degrees Fahrenheit (1.7 degrees Celsius) above current levels. In fact, if the concentrations of carbon dioxide were to grow by 50 percent, the scientists at the Meteorological Office Hadley Centre in the U.K. say the Amazon rainforest could disappear.

Luckily, temperature changes won't fluctuate much in regions around the equator in comparison to northern regions, which we talk about in the following section. How much these tropical areas will warm is uncertain.

Boreal

Boreal forests are named for the aurora borealis; these vast coniferous forests ring the northern region of the globe, from Scandinavia (where it's called *taiga*) to Russia, Alaska, and much of Canada. Boreal forests are found in northern and high Alpine regions — think pine, spruce, and fir trees.

Boreal forests will undergo many small-scale shifts while the climate changes:

- Changes in the numbers of species living in the forest
- Faster maturation and shorter life spans for trees
- Shifting relationships within the ecosystem

Rainfall patterns are also expected to change, and in many cases decrease, bringing more droughts. Taken together, these changes will add up to a major transformation in temperate forests.

Because high latitudes are experiencing great temperature changes, these boreal forests have a greater need for adaptation than tropical forests.

More forest fires

Boreal forests are naturally a fire-driven ecosystem — fires are a normal, dynamic part of the way the ecosystem currently functions. Fires clear out older trees and enable saplings to flourish. Such fires haven't been a regular occurrence, however, and they haven't spread like . . . well, wildfire. But already increased numbers of fires are affecting larger areas of forest thanks to dryer conditions brought on by human-triggered climate change.

Increased pests

Forest pests have already increased. Usually, colder temperatures keep insect populations down — most don't survive the cold winter months. With warming temperatures, however, more insects are making it through the winter. Also, certain kinds of beetles attack old or weak trees — and forests in British Columbia, Canada, are filled with old lodgepole pine. Warming temperatures combined with the age structure of the trees have led to the pine beetle populations exploding.

Mountain pine beetles in interior British Columbia have killed an area of forest as large as two Swedens. Millions of trees have been killed — trees that are the core of the province's logging economy. The lodgepole pine stands, habitat to mountain caribou and an important economic mainstay, may soon be a thing of the past. The large stands of dry, dead trees left behind are much more susceptible to forest fires. The provincial government in British Columbia expects this high population of pine beetles to continue until an early cold snap occurs, which could kill beetle larvae before they're old enough to survive winter. A similar phenomenon is happening in Norway with the spruce bark beetle.

Preparing for Mass Extinctions

When the forces acting on an ecosystem change dramatically, many of the species that live in that ecosystem are at risk.

Although scientists aren't certain how ecosystems will adapt to global warming, and even though each species and region will react differently, the Intergovernmental Panel on Climate Change (IPCC) projects that 18 to 24 percent of plant and animal species will go extinct if the Earth experiences an average global temperature rise of 2.9 to 4.1 degrees Fahrenheit (1.6 to 2.3 degrees Celsius) above 1850 levels — or 1.4 to 2.5 degrees Fahrenheit (0.8 to 1.4 degrees Celsius) above current levels. This extinction estimate keeps increasing while the temperature does, rising to 35 percent extinction with average global temperatures rising 2.9 degrees Fahrenheit (1.6 degrees Celsius) above 1850 levels — or 1.4 degrees Fahrenheit (0.8 degrees Celsius) above current levels.

Extinctions are irreversible. When those creatures are gone, they're gone!

Overall, species that migrate are expected to be the most vulnerable because their livelihood depends on the climate of the seasons. Migratory species include birds that fly south for the winter and large mammals, such as caribou, that migrate to find food. These animals are at risk because the season in which their food is available may shift more quickly than they can adapt. Species may miss feeding seasons all together. The seasons in the regions between which they migrate may no longer align. For example, red-wing blackbirds might arrive back in their northern homes to find that the marshy areas where they traditionally nest in spring have dried up. The Arctic Climate Impact Assessment by the Arctic Council expects that migratory birds will lose half of their breeding area sometime within this century.

Food sources might increase for some animals because of warmer growing temperatures, but other species, such as many kinds of birds, will suffer, when the area of their habitat declines. See Figure 8-2 for different effects that the temperature change will have on wildlife.

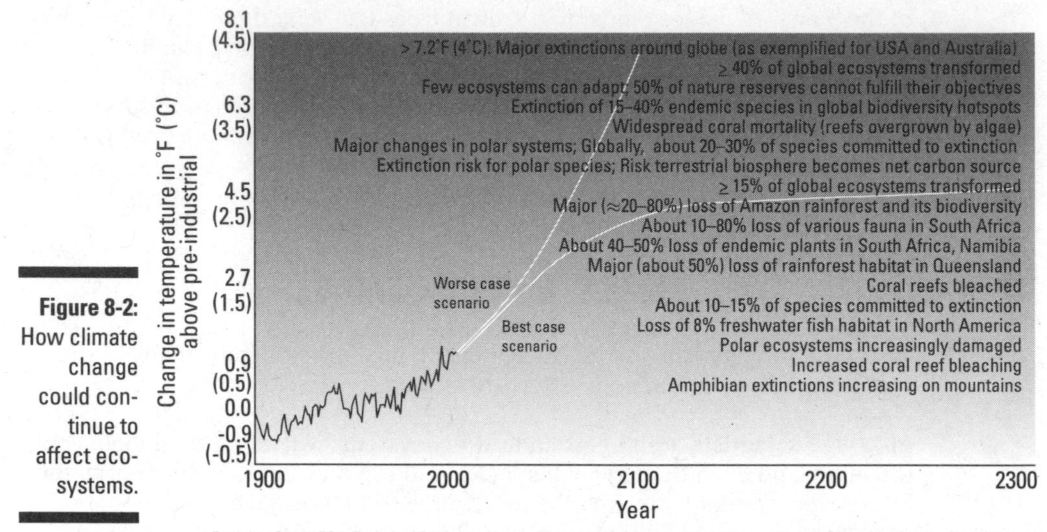

Figure 8-2: How climate change could continue to affect ecosystems.

Based on Figure TS.6, Technical Summary. Climate Change 2007: Impacts, Adaptation and Vulnerability. Fourth Assessment Report. *IPCC. Cambridge University Press.*

Life's no beach: Endangered tropical species

Plants and animals in tropical mountain regions are vulnerable to *water stress,* meaning that they either have too much or too little water. Reports from around the world show that warming temperatures affect a number of species in a variety of ways:

- **Small mammals:** Animals that have small populations (and thus a small gene pool) or small habitat ranges are at risk — a handful of these are in Australia, such as possums, bandicoots, and wallaroos.
- **Water birds:** More frequent instances of drought and lowered water tables put birds such as the Baikal teal in Asia at risk because they depend on water areas, such as marshes, to nest and breed.
- **Amphibians and reptiles:** The likes of frogs and lizards could face extinction with the amount of warming that has already happened. Disease outbreaks are occurring because climates are becoming more favorable for those diseases.

Species living in deserts (yes, things can live in the desert!) are at risk because they have to move farther to find a new suitable climate. Areas at high elevations differ greatly from land at sea level in humidity, precipitation, and temperature because of the difference in altitude, so species don't have to look far for a new home. Deserts are flat, however, and pretty much the same across the entire region. For example, 2,800 of the plant species in the

Succulent Karoo desert of South Africa will be extinct if temperatures rise 2.7 to 4.9 degrees Fahrenheit (1.5 to 2.7 degrees Celsius) above 1850 temperatures — or 1.3 to 3.4 degrees Fahrenheit (0.7 to 1.9 degrees Celsius) above today's temperatures — because they won't be able to move far and fast enough to find a livable climate. For the same reason, the Jico deer mice and pocket gophers that live in Mexican deserts likely won't be able to adapt to climate change.

Thin ice: Polar bears and other polar animals

Species in polar regions are the most vulnerable in the world to climate change. Warmer temperatures at the north and south poles will decrease ice cover, increase the temperatures of the water, reduce snow cover, and thaw what used to be permanently frozen ground. The rate of climate change may be too fast for plants to move toward the poles. These changes will put many polar species at risk of extinction, including polar bears, seals, and various birds.

Changes in the Arctic

Because the polar bear is the first large mammal that's facing extinction from global warming, it's the iconic image for climate change. Scientists estimate that 20,000 to 25,000 polar bears are left in the world. Polar bears feed on seals, which spend a lot of their time on the ice. Less ice coverage means fewer hunting opportunities for polar bears. Their lives are entirely dependent on the sea ice.

The melting ice is also bad news for the seals, even if it does reduce their chances of being claimed by a polar bear. All Arctic seals that depend on the ice for resting, breeding, and giving birth — including the bearded, ribbon, and ringed seals — are being put at risk by the ice melt.

In northern Canada, several species will be affected by the shrinking Arctic tundra, which is expected to shrink to a third of its original land cover while temperatures continue to warm. Because they'll be losing their primary source of food — species such as tundra plants — caribou and muskoxen are at risk. The Arctic tundra is also a breeding ground for geese, shorebirds, and the Siberian Crane. When the tundra shrinks, so does the area available for breeding, reducing the chance of breeding, as well.

Extinction in Antarctica

Heading down south to the Antarctic continent — known as the last great wilderness, where humans have seldom been — many species are already at risk:

- Crabeater seals don't depend on crabs to survive — despite the name. They depend on the ice, just like Arctic seals, for resting, breeding, and giving birth. As ice conditions in the Antarctic change, these seals are affected, too.
- Emperor penguins, the movie stars of *March of the Penguins* and *Happy Feet,* eat krill as part of their daily diet. With warmer waters and fewer krill, Emperor penguins have a smaller food source. Warming temperatures make for better breeding because fewer chicks die from extreme cold, but this benefit will be small compared to the loss of krill as a food source, meaning more penguins with less food.

Zoë was lucky enough to stand among thousands of Adélie penguins and their hatching chicks on an Antarctic expedition in 2007. Wildlife specialists on her expedition explained that precipitation is increasing over many parts of the Western Peninsula, where Adélie penguins thrive. So, the peninsula experiences more snow cover more often, even if it melts fairly quickly. These penguins need snow-free surfaces for hatching chicks, so this increased snowfall is a problem for them. Adélies could become extinct in some areas of Antarctica over the next 7 to 30 years.

Starting things off at the right temperatures

Climate change will benefit some species in the short term. Regions suffering from drought may have more frequent rainfall, and plants that used to be stunted in growth by chilly temperatures may now grow faster while temperatures warm. Here are a few of the short-term benefits species will see, thanks to global warming:

- Fewer deaths from exposure to extreme cold in northern countries. (This benefit will be offset by the increase in deaths from exposure to extreme heat in other areas, however.)
- Higher crop yields in northern latitude regions and greater forest growth in northern Europe. (This benefit will eventually be outweighed by negative impacts of the heat on plant health.)
- Warmer waters or land temperatures will allow some species to thrive because the greenhouse effect initially causes plants (food sources for many) to be more productive, absorbing more carbon.

The long-term benefits are highly uncertain. The IPCC estimates that costs and benefits will be roughly equal if the Earth sticks to a global temperature increase of less than 3.1 to 6.7 degrees Fahrenheit (1.7 to 3.7 degrees Celsius) above 1850 levels — or less than 1.6 to 5.2 degrees Fahrenheit (0.9 to 2.9 degrees Celsius) above today's levels. If the world goes 4.9 to 6.7 degrees Fahrenheit (2.7 to 3.7 degrees Celsius) above 1850 levels — 3.4 to 5.2 degrees Fahrenheit (1.9 to 2.9 degrees Celsius) above current levels — then the balance tips — meaning increasingly more costs and fewer benefits to species.

Chapter 9

Hitting Home: Global Warming's Direct Effect on People

In This Chapter

- Finding out where health risks exist
- Looking at how climate change affects farming, for better or worse
- Totaling the cost of global warming
- Considering how some people will feel the effects more than others

So far, global warming hasn't had a huge impact on most people's lives. The first human victims of climate change are the Inuit of the Arctic, whose traditional world is increasingly unstable because of melting permafrost and disappearing ice; the residents of low-lying islands, some of whom have already been forced to move because the rising sea engulfed their homes; and those hit hardest by increasingly violent weather events. These people are trying very hard to get the attention of the rest of humanity. But the vast majority of humans continue their day-to-day lives, oblivious to the threat.

The impact of global warming will increase in the coming years, but the degree of change will vary greatly, depending on where you live and depending on how rapidly nations around the world reduce greenhouse emissions. In some areas, farms and crops will benefit, but in others they will suffer. If you live in a northern country, warm weather might be a blessing, but higher temperatures could be a strain if you live in a southern country and have no access to air conditioning. No matter where you live, the unchecked impacts of climate change are potentially catastrophic in the long-term.

Wealth is another factor that will determine the impact that global warming has on people. For those with the resources to shield themselves, climate change will be only a costly inconvenience, at first. People without those resources won't be so lucky. But those with money will be able to pay their way out of problems at first — whether it means building new infrastructures to protect people from natural disasters or paying for importing water in the instance of drought. Unless humanity acts fast, both the rich and the poor will suffer eventually. Everyone is in the same boat.

In this chapter, we look at how changes in climate can affect your health, food, and city. We also consider how some people — such as northern indigenous communities and women around the world — will be particularly adversely affected by global warming.

Like all the information in this part, this chapter may get you down. But remember, these projections merely suggest what might happen if civilization stays its current course. People can stop the impact of global warming from becoming devastating — and some effects can be avoided entirely — but more people must start cutting our carbon emissions right now. Check out Parts IV and V for more about how civilization can avoid the worst case scenarios that we discuss in this chapter.

Health Scare: Outbreaks and Diseases

Few scientific studies have been conducted to examine the impact global warming alone will have on people's health. From examining what little information they have, the Intergovernmental Panel on Climate Change (IPCC) states that climate change is adding to the number of people suffering from disease or early death. The World Health Organization estimates that climate change was already responsible for 150,000 individual deaths around the world in the year 2000. Because global warming's effects have worsened since that time, that number has probably risen.

Climate change will affect health issues in both positive and negative ways. The bad impacts on health will, unfortunately, greatly outweigh the good. Figure 9-1 gives an idea of the intensity of these impacts, weighting the negative impacts against the positive. *Very high confidence* translates into a minimum 95 percent certainty; *high confidence* is 60 to 95 percent certainty, while *medium confidence* is 40 to 60 percent certainty.

Malaria

The link between *malaria* (a disease carried by mosquitoes and causing chills and fever that can lead to death) and climate change is unclear. About five years ago, scientists argued that climate change could cause malaria to spread. Now, however, scientists believe that occurrences of malaria will vary by region depending on the climate changes of the region. For example, malaria transmission will likely decrease in southeastern Africa in the next 12 years, but the risk of malaria will increase in industrialized countries including Australia and the UK.

Two climate conditions can make a region a mosquito hotbed: high rainfall and warm temperatures. Still water and warm weather make for a perfect romantic getaway for mosquitoes. Rainfall can leave puddles and pools of water, still and ready for action. Warm temperatures allow mosquitoes to survive. Regions with this type of climate, or that will be experiencing this type of climate in the near future, may be more susceptible to malaria if mosquitoes in that region are affected.

Many other factors contribute to malaria transmission, such as the national health infrastructure and education, so definitively identifying climate change as a cause of the disease's spread is almost impossible. For example, although malaria has returned to the highlands of Kenya, scientists don't know why it has. Some say it's because of warming temperatures, but it could just as well be that the disease has developed a resistance to anti-malarial drugs or because of some other environmental changes.

	Negative impact	Positive impact
Very high confidence		
Malaria: contraction and expansion, changes in transmission season	←	→
High confidence		
Increase in malnutrition	←	
Increase in the number of people suffering from deaths, disease, and injuries from extreme weather events	←	
Increase in the frequency of cardio-respiratory diseases from changes in air quality	←	
Change in the range of infectious disease vectors	←	→
Reduction of cold-related deaths		→

Figure 9-1: The projected positive and negative climate change impacts on human health.

Based on Figure 8.3, Chapter 8: Human Health. Climate Change 2007: Impacts, Adaptation and Vulnerability. Fourth Assessment Report. *IPCC. Cambridge University Press.*

Scientists do know, however, that the range of malaria-borne mosquitoes has the potential to shift because of increased climate changes around the world. Those mosquitoes may leave some areas and move into others.

Cholera

Global warming doesn't directly cause *cholera,* an infectious disease that causes major cramps and diarrhea, but its effects may create an environment in which the disease can flourish. Cholera isn't fatal if treated promptly, but it can be very dangerous in parts of the world with shaky public health systems.

Cholera usually happens in areas with bad or non-existent sewer systems, where human excrement ends up mingling with drinking water. Flooding brought on by climate change might lead to a cholera outbreak in those regions because flood waters can wash human and animal waste into sources of drinking water. South Asia, for example, could have an increased risk to and toxicity of cholera if local ocean water temperatures rise, reports the IPCC. (Refer to Chapter 7 for more about the risks of flooding.)

Other problems worsened by global warming

The IPCC points out a number of health concerns that civilization already deals with that might worsen because of climate changes and extreme weather events:

- **Allergies:** If you have allergies to pollen and dust, those allergies could get worse, depending on where you live. If global warming brings you an early spring, that early spring will bring pollen, too — extending your allergy season. Countries such as Canada, Finland, and the Netherlands will probably be most affected by the increase in pollen because they're undergoing larger seasonal changes than more southern countries.
- **Contaminated drinking water:** This could be a risk for areas that see an increase in rainfall and flooding because of global warming. During Hurricane Katrina in the U.S., for example, water supplies became contaminated, and many cases of diarrheal diseases appeared, some of which were fatal. (We talk about flooding and water contamination in detail in Chapter 7.)
- **Dengue fever:** Found in warm climates, this disease causes severe joint pain. Like malaria, it's carried by mosquitoes. A pool of still water is a mosquito's honeymoon suite. Consequently, a lot of rain (creating more pools of water where mosquitoes breed) combined with temperatures

warm enough for the mosquito to survive can increase its spread — but oddly, so can drought. When drought hits an area, more people store water outdoors, creating other excellent mosquito-breeding sites. The IPCC expects that global warming will create climates in both New Zealand and Australia that are favorable for mosquitoes and for dengue, and that both countries will see more species of mosquitoes that can carry the virus.

One report shows that about a third of the world already has favorable conditions for dengue and that 5 to 6 billion people will be at risk of dengue by 2085, compared to the 3.5 billion that would be at risk without climate change as a factor.

- **Diarrheal disease:** Think extreme diarrhea. This disease occurs most frequently in Australia, Peru, Israel, and islands in the Pacific Ocean when the temperatures soar and rainfall patterns change. Bacteria thrive in higher temperatures, and flooding increases the risk of infecting drinking water sources. Areas that have poor sanitation have even greater risk.
- **Lung problems:** Air pollution often worsens health issues, such as asthma. Smog episodes, for example, are more intense during heat waves — so, you find more lung problems in big cities than anywhere else. Many greenhouse gases — methane and nitrous oxide to name two — just happen to also pollute the air. The women and children in some developing countries have a higher prevalence of lung cancer because of the smoke from fires that they burn to cook their meals — whereas the men aren't as affected since they are out of the house for most of the day.
- **Lyme disease:** Carried by ticks that hang out in trees and burrow into your skin (yuck), Lyme disease can cause everything from joint swelling, fever, and a rash to significant disability. For example, Lyme disease has moved north in Sweden because of milder winters and is becoming more common in Portugal and the Netherlands. It has spread in North America, as well.
- **Skin cancer:** Too much direct sun exposure can cause this form of cancer. In areas that experience rising temperatures, people often want to be outside in the sun more and wear less clothing. To add to the problem, greenhouse gases (GHGs) actually cool the upper layers of the atmosphere, which creates perfect conditions for ozone-depleting gases that create holes in the ozone layer. And that means less protection from the sun. An increased risk of skin cancer is particularly worrying in places with depleted ozone, such as Australia, where people already have a high risk to dangerous sun exposure.
- **Vermin:** Rats, mice, and other rodents often carry disease. Flooding or heavy rain pushes these little critters out of their burrows and directly into the paths of humans. Low-income countries that are susceptible to flooding, such as many within Central America, are particularly at risk.

Putting Pressure on the Fields

Because agriculture depends on the land and the climate for its products, it'll feel the impact of global warming more than any other industry, which could cause increased inconsistency and uncertainty for the world's food supply. The temperature changes, severe storms and rains, and altering growing seasons that we discuss in Chapter 7 will affect farmers worldwide while climate change develops. But those changes won't affect all farmers in the same way.

Affecting farmers

Agricultural productivity could increase or decrease, depending on the region. Northern Europe, for example, should have better growing seasons, but areas around the Mediterranean probably won't do as well.

Currently, more carbon dioxide in the atmosphere and longer growing seasons are helping some crops grow. Northern Hemisphere regions, such as southern Canada, are already seeing benefits from warmer temperatures — a 1.8 to 5.4 degrees Fahrenheit (1 to 3 degree Celsius) boost in local temperatures above 1850 levels, or 0.4 to 4 degrees Fahrenheit (0.2 to 2.2 degrees Celsius) above current temperatures, means a boost in crop yields in these places. Plus, farmers can plant many crops sooner when the seasons shift. The International Food Policy Research Institute expects a 56-percent increase in world cereal crops and a 90-percent jump in livestock between 1997 and 2050 in regions such as the U.S. and Canada, where an adequate water supply exists. Crop increase beyond 2050, however, is not likely.

On the other hand, regions such as southern Africa will suffer from an increasingly warm and dry climate. There, rainfall isn't only becoming less frequent, but also less predictable. These conditions together are actually reducing the length of the planting season in southern Africa, the opposite of what's happening in the Northern Hemisphere. Tropical regions will likely produce far less food from crops, especially rice.

In the long-term, the IPCC expects that the overall effects would be negative around the world. If people allow temperatures to continue to increase, they could exceed what many crops can stand, and those crops won't survive. Some crops in dryer or lower-latitude regions suffer from even a small temperature rise. For example, mango and cotton crops in Peru have a shorter growing season when temperatures are higher than normal. Temperature aside, extreme weather events, such as heavy rains or droughts, can stress crops by either drowning or dehydrating them, enough to upset any farmer depending on selling those crops. Pest outbreaks can damage crops, and pest outbreaks become more frequent when the climate becomes more favorable to them. These outbreaks are not easy to project because all pests are different and climates differ regionally. To prepare for these possibilities, *agronomists* (scientists focused on crops and soil) are trying to develop crops that are more pest- and drought-resistant.

Hurting the global food supply

First, the good news: The IPCC predicts that the number of undernourished people in the world will decrease dramatically in the next few decades, thanks to quickly developing countries and their growing economies.

Now, the bad news: If people don't fight global warming, progress will be stalled and that decrease will occur at a much slower rate. About 820 million people in the world are undernourished. By 2080, the number of malnourished people could decline at least to 230 million — not quite making poverty history, but definitely making an improvement. With climate change, civilization could make less progress, ending up with 380 million malnourished people instead.

If global warming increases unchecked, the number of hungry could be almost 40-percent higher than if humanity addresses climate change. And some estimates in IPCC reports say that it could take 35 more years to cut world hunger in half if civilization doesn't act on climate change; addressing climate change could see the number of undernourished people in the world halved by 2015, but that goal won't be reached until 2050 if emissions continue to grow and temperatures continue to rise.

The IPCC reports that world food production will likely grow when local temperatures rise 1.8 to 5.4 degrees Fahrenheit (1 to 3 degrees Celsius) above 1850 levels — or 0.4 to 4 degrees Fahrenheit (0.2 to 2.2 degrees Celsius) above current temperatures — but these numbers don't apply everywhere equally. In both dry and tropical southern regions of the globe, the IPCC expects crop production to drop when just a 1.8- to 3.6-degree Fahrenheit (1- to 2-degree Celsius) local temperature shift occurs. This drop in crop production will then directly increase that region's risk of hunger. And globally, with any increase above 5.4 degrees Fahrenheit (3 degrees Celsius), major crops of corn and wheat will be below normal in low latitude countries, such as Brazil and Kenya, whether farmers try to adapt or not. This amount of warming will also stress and even kill livestock in semi-arid areas such as inland eastern Australia or dryer regions of Texas. For agriculture in northern countries, such as Canada, the IPCC reports that temperatures are rising more than the world average — the Arctic, for example, has warmed almost twice as fast as the rest of the world over the last century.

Warmer weather: A gift to wine drinkers?

Recent research has shown that warming weather has boosted wine production. Just as vineyards and wineries thrived during the medieval warm period, the IPCC reports that the climate in northern Europe and in parts of the United States, such as California and Oregon, is much more favorable for growing grapes than it was previously. Before you get your hopes up, remember that climate change also means more extreme and varied weather, not just a nice warming touch. In the spring of 2006, Californians woke up to icicles hanging from their fruit trees. And Canadian ice wines, which are made from grapes that they first freeze on the vine, have been suffering from warmer temperatures.

Paying the Price for Global Warming

Debate over global warming really heats up when it comes to money. Because industry contributes a lot of the excess greenhouse gases in the atmosphere, some governments, such as the United States under President George W. Bush, worry that cutting greenhouse gases would have an adverse effect on the economy.

To be honest, reducing greenhouse gas emissions will have a financial impact. In 2006, Sir Nicholas Stern, former Senior Economist to the World Bank, reviewed the economic impacts of the climate crisis at the request of the U.K. government. His report, "The Stern Review on the Economics of Climate Change," looked at the financial impact that global warming would have on the world's economy. In his most recent report, "Key Elements of a Global Deal on Climate Change," he calculated that acting now to reduce greenhouse gas emissions would cost the world a cut of 2 percent of global *gross domestic product* (also known as GDP, the global measuring stick of economic wealth) annually over the next several decades. India already spends 2 percent of its GDP on adapting to climate change impacts.

Stern's report also examined the cost to the world if people did nothing and greenhouse gas emissions weren't reduced. If humanity doesn't reduce greenhouse gas emissions now, the report found it will cost Earth's population five times more than if it does, resulting in a cut of 5 percent of the GDP — every year. And that's one of the better-case scenarios. The worst-case scenarios show that waiting to reduce emissions could cost the world 20 percent of global GDP or more. Failing to reduce greenhouse gases could cost the world economy $7 trillion!

Despite the projections in the Stern report, the GDP has continued to grow an average of 34 percent in industrialized countries, even as those countries cut greenhouse gas emissions (by 3.3 percent between 1990 and 2004, for example). Germany cut greenhouse gas emissions by 17.2 percent while watching its GDP rise 28.6 percent in that time frame.

Many European countries have benefited with continued GDP growth because they started acting on climate change decades ago. The Stern reports give a clear message: The longer humanity waits to act on climate change, the more serious the impacts become, and the more it costs humanity to adapt and recover.

Highways, waterworks, and the other stuff humans build

In Chapter 8, we talk about natural systems at risk because of climate change, but human-built systems are at risk, too. Governments call it *infrastructure* — such as the roads that you drive on and the waterworks that take away sewage and deliver *potable* (suitable for drinking) water. It's the stuff that humans have built to make modern life easier. And civilization built all of this stuff for the climate it used to have.

No matter where you are in the world, the costs from extreme weather — major flooding, fires, landslides, and storms — have been increasing for the last three decades. These natural catastrophes, brought on by civilization's unnatural addition of greenhouse gases to the atmosphere, endanger the infrastructure that's the backbone of our cities. The repair bill for the following damages won't be cheap:

- **Buildings:** Storms and flooding can quickly damage unstable buildings. People living in inadequate housing that is easily damaged by strong winds or storms are especially at risk from the extreme effects of climate change.
- **Electricity demand:** Hotter days mean cooler buildings when people run their air conditioners to the max. Heat waves come hand in hand with skyrocketing electricity demand and, often, major blackouts.
- **Sewage systems:** Storms and flooding can also cause the sewage systems in cities to overflow.
- **Transmission lines:** You've likely seen downed telephone poles and power lines after a big storm. Damaged transmission lines could become a more common sight for many parts of the world.
- **Transportation:** Highways, roads, and railroad lines will all require more frequent maintenance and repair when they're subjected to extreme weather.

Although everyone will be affected by the physical impacts of climate change, some people will be more vulnerable to negative changes. The poor are particularly at risk. Poverty, combined with a lack of social support, was the main cause of heat wave deaths in the Chicago heat wave of 1995. People in coastal cities — accounting for 10 to 23 percent of the world's population — are vulnerable to sea level rise and flooding from storm surges because of their proximity to the ocean. Cities that don't have much green space will be at greater risk, too; without soil and trees that stabilize the ground and absorb water, these concrete cities are more vulnerable to landslides and flooding.

An unfair split: Costs to the industrialized and developing nations

Industrialized countries, such as the U.S. and the U.K., have pumped most of the excess greenhouse gas emissions into the atmosphere — through industry; their heavy reliance on cars; and their ever-growing, ravenous consumption of electricity and energy. Developing countries don't have the amenities that create greenhouse gas emissions. The effects of global warming don't care who's to blame, however. Because developing countries have fewer resources and less financial ability to recover from natural disasters, they will feel the first major impacts of climate change.

Developing countries don't have the economic resources to adapt to climate change and to recover from its worst effects. The U.S. is still cleaning up the damage caused in New Orleans by Hurricane Katrina in 2005, with a final bill that some say might be as much as $150 billion. Katrina happened to one of the wealthiest nations on Earth. Imagine the impact for a nation such as Honduras or Guatemala if Katrina had made landfall there. The IPCC reports that in 1985 and 1999, for example, natural disasters lost about 2.5 percent of the gross domestic product (GDP) of the world's richest nations, and the poorest nations suffered a loss of 13.4 percent of their combined GDP.

The financial costs that come along with extreme weather events — storms, flooding, and droughts — give any country a huge economic hit. Because they lack the industrialized countries' strong, diversified economies, developing nations have far fewer economic resources to bounce back from such a huge disaster — yet they're the most likely to be hit, and hit first, because of their geographical locations. Ninety percent of deaths caused by natural disasters happen in developing countries. You've likely seen stories in the news about these natural-disaster deaths, such as the 30,000 people killed from flash floods and landslides in Caracas in 1999, or the 15,000 homes damaged by major flooding in Cape Town in 2001. Though climatologists can't link these events directly to climate change, the trend is consistent with climatic events. They also demonstrate that developing regions are more vulnerable to extreme weather than industrialized areas.

Developing nations need to move quickly to adapt — with measures ranging from regulating freshwater use to building levies to planting forests — to cushion the inevitable impacts of climate change. (We discuss how developing nations can adapt to global warming in Chapter 12.) The money that those countries are spending on adaptation was originally earmarked for further development. Consequently, development is slowing while adaptation measures are growing. The Stern Review Report on the Economics of Climate Change expects adaptation to cost tens of billions of dollars annually, just for developing countries.

While countries around the world begin to apply solutions to climate change, developing countries will probably get partners to support their adaptation measures. Carbon markets and country-to-country partnerships within the Kyoto Protocol (such as the Clean Development Mechanism) are two major tools that can help alleviate pressures on developing nations. We talk about these solution-based projects in Chapter 12.

Feeling the Heat First: Unequal Effects

Just as climate change affects regions differently, it affects groups of people differently as well. Unfortunately, the impact of climate change will be most keenly felt by those who have few resources to adapt. Activists who seek to address the imbalance of who's causing climate change and who's being affected by it refer to their cause as *climate justice.* In this section, we look at some of the major climate injustices various populations around the world are experiencing — and will experience.

Heavy warnings: Beyond the IPCC

Not too many years ago, the U.S. Department of Defense (a surprising source) studied the risks of climate change to global security. The report, titled "An Abrupt Climate Change Scenario and Its Implications for United States National Security," was covered in *Fortune* magazine in 2004. The Pentagon study concluded that the risks of climate change were more significant than the risks of terrorism.

Uphill battle for downhill skiing

You know something's a big deal when it needs an international conference. The year 2003 saw the very first International Conference on Climate Change and Tourism.

Ski hills around the world have been closing in recent years because of rising temperatures and too little snowfall. The ski hills that are still operating face shortened ski seasons and make do with machines that make snow.

Ice skating, outdoor hockey, downhill and cross-country skiing, snowboarding, the luge, sledding, snowshoeing, and extreme sports such as ice climbing are all at risk.

Climate change won't hurt just cold weather tourism — islands that lure people to the beaches each winter need to be thinking about rising sea levels and storm surges, and how those changes will affect their infrastructure. Coral-reef tourism could change or even cease, and the tourist season in certain countries could shift.

Northern communities

The north is seeing some of the strongest changes brought on by global warming, and these changes are threatening the way of life for many communities.

Indigenous communities are particularly affected. The permafrost and thick winter ice that served as their road surfaces are melting. Traditional food sources are becoming hard to obtain; the melting ice makes fishing and hunting difficult. Thin ice has caused several hunting accidents.

Climate change is making the weather less predictable than ever before. Many people think about the weather only when they're going somewhere or heading outside for leisure or recreation. But in these indigenous communities, weather helps define their way of life.

Some northern communities think they can fully adapt to all the changes. Others are considering moving to new locations where they can live more easily, and still others are enjoying longer hunting seasons. But all these communities are experiencing noticeable change, and it's not slowing down any time soon.

Changing culture and lives

The Inuit have been adapting to these changes, but at the cost of their traditional culture. The Inuit are a hunting people, but they now have shorter and shorter hunting seasons because the ice melts earlier and earlier each year.

While their environment changes so rapidly, knowledge that was passed down for generations becomes unreliable. In an interview with the *Nunatsiaq News,* Naalak Nappalak, an elder from Kangiqsujuaq, described how he can no longer confidently predict the weather: "Before we knew by looking at the sky whether there would be storms or if it would be calm. Nowadays just when you think you know how the weather will be, they can change in an instant. It's this inconsistency that is most noticeable."

People living outside of the native village of Kotzebue in northwestern Alaska can travel into the town for supplies only when the ice is sufficiently frozen and stable for travel. Warming temperatures mean a longer thawed season, which means people who need supplies or medical attention have longer to wait. The ice also usually serves as a barrier to storm surges, and less freeze-up over the year can lead to erosion and flooding along beachside roads.

People living in the indigenous community of Lovozero in Russia have had to deal with climate change in a number of ways. As one local described it, "Bogs and marshes do not freeze immediately, rhythms change, and we have to change our routes of movement, and this means a whole new system of living is under change. Everything has become more difficult."

Affecting the hunt

Some northerners can hunt more easily because of climate change, however, including those who fish for larger whitefish, as well as clamming harvesters and seal hunters (because shorter periods of sea ice give longer water-based hunting periods). Access to wood for fire fuel has improved, too, because it washes up on shore for a longer period of time each year.

Although northerners may hunt more easily in the short-term, climate change threatens the survival of many species (as we talk about in Chapter 8). Animal populations, such as the seals that depend on sea ice, can become easily stressed by global warming, which translates into stress put on humans in the long-run.

In Qaanaaq, Greenland, people describe that just seven or eight years ago, they used to be able to go out onto the ice to hunt as early as October. Now, these hunting grounds sometimes don't freeze up until January. Temperature, coupled with changing sea currents and wind conditions, causes these changes. The places in this area of Greenland are named for their natural and geographic characteristics. Ironically, some of these places no longer fit their names — *Sermiarsussuaq,* which means "the smaller large glacier," used to cover the landscape all the way to the ocean's edge. It no longer exists.

People in poverty

Poorer populations in developing and industrialized countries are very vulnerable to the effects of climate change. Poor people don't have the resources to adapt to extreme changes — especially unexpected ones.

Half of the world's population earns less than $2 per day and largely depends on public services. In some developing countries, public services may not be able to cope with the aftermath of extreme weather events brought on by global warming. In other countries, public services don't even exist, leaving the poor with few resources to cope. As the IPCC says, "This does not necessarily mean that 'the poor are lost'; they have other coping mechanisms, but climate change might go beyond what traditional coping mechanisms can handle."

Women

You've likely heard or read about women's rights and women's equality issues. But you probably don't often hear about women's inequality when it comes to global warming. The people working closely on the climate change issues are only now giving it attention.

You may not realize it, but your *gender* (your sex and how it influences the roles you play in society) affects the degree to which global warming may hurt you. In both rich and poor countries, women tend to bear the brunt of the climate change's negative impacts, mostly because they tend to be whole lot poorer. In fact, the World Conservation Union (IUCN) reports that 70 percent of the world's poor are women.

Women in developing nations

In many developing countries, women are frequently agricultural workers. They're the main family caretakers responsible for producing food for the family. The IUCN reports that women produce 70 to 80 percent of the food in sub-Saharan Africa. Latin America isn't far behind at 65 percent. Scientists predict that these poorest regions are also the ones to be hardest hit by climate change, which we discuss in the section, "Paying the Price for Global Warming," earlier in this chapter. When climate change negatively affects a region, men often migrate to cities to find better paying jobs, leaving the women and children behind to try to survive on the land, which is no longer productive because of climate impacts such as drought.

How natural disasters caused by climate change affect women

Natural disasters, which will increase in the wake of climate change, affect women and men differently. Women, as main caregivers, are more likely to be indoors — particularly in developing countries — when a disaster occurs and won't be able to escape. Even if they do survive, women tend to stay within the community longer afterwards to care for their families, thus exposing themselves to deadly diseases.

Although not linked to global warming, the grave impact that natural disasters have on women can be seen in the death toll from the major Asian tsunami that struck at the end of 2005 and hit the province of Aceh in Indonesia, where 75 percent of those who died were women.

When the death toll from natural disasters has significant gender differences, the resulting gender imbalance in the society can have major, long-term negative consequences. The Asian tsunami left the society with a three-to-one ratio of males to females. With so many mothers gone, the area experienced increases in sexual assaults, prostitution, and a lack of education for girls. Research in this area is still in its infancy, but the IPCC has reported that women are more likely than men to suffer from post-traumatic stress disorders after living through a disaster and that men are likely to commit domestic violence against women after natural disasters. This is worrying because climate change is expected to increase the intensity and frequency of storms and extreme weather events around the world.

Gotelind Alber: Getting gender on the agenda

Gotelind Alber, a German physicist who's worked in the energy and climate policy sector for 20 years, is bringing gender issues into the discussion on climate change. Whether working in the fields of science, technology, climate change, or policy development, she has always worked to put gender on the agenda. She co-founded the international network Women for Climate Justice.

Among her long list of projects is the "Climate for Change: Gender equality and climate policy" project. The goal of this European-based initiative is to

- Help balance the participation of women and men in creating climate-related policies.
- Address the fact that major climate change–related sectors (such as energy, transportation, and buildings) are mostly run by males.

This project's aim is to stress the importance of a balanced gender view in decision making. They're working with ten European cities to integrate a gender perspective into their climate change policies. More information and the ongoing results of this project can be found at www.climateforchange.net.

Part IV
Political Progress: Fighting Global Warming Nationally and Internationally

"We post these at the climate change conferences. I like to think they send a subtle message to the leaders."

In This Part . . .

Governments are on the frontlines of the fight against global warming. In this part, we investigate what governments can do at every level, from your local mayor to the leader of your country. Because climate change is a global matter, national leaders need to collaborate, as well; we look at how countries are banding together. One group of nations, however, is particularly challenged by global warming; in the last chapter in this part, we look at how developing nations can begin to cope with this problem.

Chapter 10

Voting for Your Future: What Governments Can Do

In This Chapter

- Getting the goods on what your government can do for the climate
- Taking a look at taxes and laws that can help fight global warming
- Identifying cities, regions, and countries that are saving money while reducing carbon

Many people can be cranky about their country's government and cynical about its motives and whether it can really make any positive changes. And no one likes to pay taxes, even if those taxes do pay for basic services that people need, such as garbage pick-up, drinkable water, schools, and so on. Governments can often seem intrusive, taking your money and telling you what to do instead of the other way around.

Ever the optimists, we like to take a more positive spin on governments. A government is the servant of the people in a democracy. In the battle against global warming, governments are invaluable allies in stopping — and reversing — climate change. The Intergovernmental Panel on Climate Change (IPCC) reports that reducing greenhouse gases as much as the planet needs requires government leadership in the form of regulations and programs.

In this chapter, we take a look at how governments can fight — and are fighting — climate change through initiating programs, and regulating and taxing emissions. Knowing what solutions governments can enact empowers you as a voter; you can knowledgably select candidates whose climate change plan seems like it'll be the most effective. Beyond that, by letting candidates know that you're voting for action on climate change, you will encourage all candidates to give this crucial issue a higher priority. We also share some success stories, so you can see what your votes can do!

The non-political political issue

Climate change has become a hot political issue. Depending on the political culture of your country, the issue is non-partisan, pan-partisan, or more than a little partisan.

In some countries — such as the U.S., Canada, and Australia — the conservative versus liberal fight on climate change has never been more heated. The defeat of John Howard's government in Australia in 2007 may well be the first election globally in which climate change was a significant factor. New Prime Minister Paul Rudd made ratifying the Kyoto Protocol one of his top priorities, delivering on his election promise within days of becoming Prime Minister. In the U.S. and Canada, parties to the right of the political spectrum have consistently opposed the Kyoto Protocol. (The situation is more complex at the state and provincial levels, however. The governments of California and British Columbia, arguably the most progressive jurisdictions in the U.S. and Canada, respectively, both have "right-wing" governments.) At the time of publication, however, things seem like they may change in the States; the U.S. Republicans, led by John McCain, favor action on the issue.

In Europe, however, the fight against global warming isn't as divided along party lines. In the U.K., the social democratic government is in power, and the conservative opposition is challenging the effectiveness of the government's efforts to address the climate crisis. The right-wing president of France, Nicolas Sarkozy, has been very progressive on the climate front. Despite some criticism of his national crush on nuclear power, he's leading the country and befriending environmentalists along the way. His inspiring speeches stress the need for major countries such as China to lead the way in low-emission development. France and China recently signed a joint statement committing to a partnership to help each other reduce harmful emissions.

Perhaps the most influential conservative leadership on climate change has been from German Chancellor Angela Merkel. She was Minister of the Environment in the early 1990s when the mandate to negotiate the Kyoto Protocol was just beginning, and Merkel surprised some observers by maintaining the previous left-leaning government's course for greenhouse gas reductions when her party came to power. She also led the European Union in the same year that she helped push for major reductions in energy use. In 2007, she led G8 discussions, keeping a focus on climate change within the G8.

If They Had a Million Dollars . . . (Wait — They Do!)

When it comes to cutting greenhouse gas emissions, money is a powerful tool. Governments can create ways that people and companies benefit financially by making the right environmental choices. These incentives can take the form of tax credits and rebates, or governments can directly fund initiatives that plan to cut greenhouse gas emissions. Governments can also tax — as you probably know all too well. Taxes flow to government coffers for public services. But tax policies can also send a message — by what gets taxed and what gets subsidized. Governments can make reducing carbon-based fuels a smart financial decision by taxing carbon.

Creating incentives

An excellent way for governments to encourage proper behavior (climate change–related or otherwise) is to offer *incentives* — a sort of reward that encourages people (or organizations) to act in a certain manner. You were probably offered incentives as a little kid to get good grades or tidy up your room. (Governments are likely to offer more than chocolate sundaes to entice people to help fight global warming, however.)

Incentives can come in many forms: a direct cash return or rebate, a tax break, or a lower consumer price thanks to subsidies to the manufacturers. Not all subsidies and incentives are announced and heralded. Some are embedded deep in a tax system and allow corporations to write off costs and reduce the taxes they have to pay.

Incentives are popular with government officials, who would rather reward good behavior than risk alienating the electorate through regulations and taxes. Observers such as the IPCC argue that incentives are most effective when governments use them in combination with regulations, which we talk about in the section "Laying Down the Law," later in this chapter. Otherwise, it's a case of too much carrot, not enough stick.

Here's a list of incentives that the IPCC recommends:

- **Agriculture:** Most governments already subsidize agriculture; to help combat climate change, governments can subsidize farmers who take active steps to reduce their greenhouse gas emissions by such measures as improving land and soil management and using efficient farming methods. (We talk about how farmers can go green in Chapter 14.)
- **Consumer goods:** Because low-emission products can sometimes cost a little more than inefficient wasteful ones, the government can encourage people to buy the more expensive product by footing some of the bill. A few examples include the following:
  - **Canada:** The government offers grants of up to $5,000 so that owners can retrofit homes to make them more energy efficient.
  - **Japan:** The Ministry of Economy, Trade, and Industry offers subsidies for homeowners buying automated systems such as thermostats and electricity regulators, which help reduce energy consumption.
  - **United States:** The federal government offers rebates to anyone buying a hybrid car.

- **Forestry:** Subsidies to this industry could increase the amount of forest space, reduce or eliminate the logging of old-growth forests (which store far more carbon than newly planted forests), and encourage forestry companies to manage harvested forests in a sustainable fashion, keeping the ecosystem and soils healthy. (Check out Chapter 2 to find out why trees and soils are important to the climate.)
- **Waste management:** National or regional governments could offer financial incentives to encourage good waste management practices within their cities and communities. These incentives could include encouraging the diversion of waste from landfills through reduced packaging and recycling. For the remaining waste, initiatives could include incentives to capture and/or use methane-containing landfill gas, or funding local purchases of technology that breaks down waste more effectively than landfills or processes the methane gas emitted from landfills into clean electricity.

Planning for emissions trading

Emissions trading, which is sometimes called *cap and trade,* is a complex and controversial approach to reducing greenhouse gases. In this system, a government decides how much carbon a geographical region (such as a city) or a sector (such as all facilities that generate electricity from coal) is allowed to emit. The allowable level is called the *cap.* The trading comes in when an emitter produces a level of carbon less than its cap — it can sell the difference to another emitter that has gone over its limit. The promise of the revenue from selling their extra credits encourages companies to accelerate the adoption of new low-pollution technologies. Over time, the government lowers the cap, gradually reducing the allowable credits. The intended result is less and less pollution.

Within North America, non-greenhouse gas emissions trading has had some historical success on other environmental issues such as air pollution. Los Angeles, for example, improved air quality through trading nitrogen oxide and *volatile organic compounds* (liquid chemicals that can immediately evaporate into gas at room temperature) credits. The province of Ontario trades nitrogen oxides and sulfur dioxide, and it's now setting up a larger cap-and-trade program with the neighboring province of Quebec.

Europe has had an active greenhouse gas market since 2005. It was the first market of its kind to span different countries and numerous sectors. The European Union (EU) carbon market is the model used around the world for building a market that trades carbon dioxide emissions and credits. This system may, in 2011, begin to include airline emissions, a major and growing piece of global carbon dioxide emissions.

Carbon trading has its upside and its downside. On the one hand, it regulates emissions measurements and enables governments to hold companies accountable for the measured emissions. On the other hand, putting a price on carbon is a thorny matter. Do governments charge for only the carbon being directly emitted by burning coal? Or do they put a price on that *and* on the emissions produced by mining out the coal and shipping it?

Emissions trading is often attacked as a "license to pollute." (Of course, in most jurisdictions in North America no restrictions whatsoever exist on emitting greenhouse gases; right now, people and industry can emit as much as they desire — for free!) It can actually end up being ineffective if the carbon price is set too low. The European Union Emissions Trading System, for example, has been in place for four years, yet countries such as Germany are still putting up new coal plants because the added cost of the cap and trade is so marginal that it doesn't influence their decisions. Market mechanisms can work, but only if the price is effective and governments monitor trading to prevent cheating and steadily reduce the cap over time.

Industries can engage in emissions trading without government involvement, too — that is, they can limit their net greenhouse gas emissions on a voluntary basis. Although the voluntary market expanded rapidly in 2005 and 2006, it is still $^{1}/_{100}$ the size of the regulatory market in greenhouse gas emissions trading. Take a look at Chapter 14 to read about the Chicago Climate Exchange, the first private emissions trading market.

Putting programs into place

One way governments can fight climate change is through engaging and empowering their population. By supporting research, educating people, and helping people adapt to the changes that global warming will bring, governments can put the tools for survival in the public's hands.

Research

Government funding for energy research skyrocketed in the mid-1970s, during the oil shock (when crude oil prices rose sharply), to encourage scientists and companies to develop alternative energy sources. But when the first crisis faded, so did interest. Government investment in energy research around the world is half of what it was in 1980. Now that the world is facing a "climate shock" and is reeling from high oil prices, governments can encourage similar initiatives.

The private sector and government researchers need funding to develop low-emission technologies at roughly the same rate that the climate is changing. *Low-emission technology* is essentially any kind of technology that reduces greenhouse gas emissions: It can be renewable energy technology, carbon capture and storage (see Chapter 13), or more efficient appliances and lighting (see Chapter 18).

The IPCC recommends that governments invest in two types of research to help reduce greenhouse gas emissions and develop renewable energy technologies:

- ***Private* research** is done by companies. The downside is that the results are owned and sometimes even patented by the company, restricting access to the new technologies they develop. The upside is that results drive the research, and that research is often in line with the market and what consumers want — in other words, it gets the job done.
- ***Public* research** is done by the government itself. The upside is that anyone can study and use the results, but public research often takes a long time because no market push drives the research.

Education and awareness

Global warming is a complex issue — you could even write a book about it! Not surprisingly, people often have a lot of misconceptions about climate change and what they can do to help reduce their carbon emissions. Because not everyone's going to read *Global Warming For Dummies,* governments can play a big role in educating people so that those people can make informed choices.

The United Kingdom's government has done an excellent job of educating its citizens about climate change. Here are a couple of the initiatives launched by the British government:

- A national climate communications campaign, dubbed Act on CO_2 (See Chapter 22 for information on this great Web resource.)
- A large-scale, climate-awareness, art exhibition, featuring world-class photographers whose pictures portray the global impacts of climate change, as well as the various global tactics taken to reduce carbon emissions

One great way for governments to raise climate change awareness is to include it in school curriculums, creating a profound effect on society when these students grow up. Finding out about climate change at an early age could influence decisions these students make for the rest of their lives.

Of course, awareness campaigns don't necessarily reduce personal greenhouse gas emissions — in fact, ample evidence demonstrates that education and exhortation on their own produce very little effect. Arguably, they do create a heightened awareness of global warming. More importantly, public awareness is essential to ensuring that the public understands and accepts other carbon-dioxide-reducing measures that the government may implement.

Adaptation

The climate is already changing, in some areas more than others. It's going to keep changing over the next century, even if humanity is successful in keeping the overall magnitude of climate change to below truly dangerous levels at the global scale. Governments can play a key role in dealing with the changes civilization can no longer prevent by providing funding and resources to help people and businesses adapt.

Global warming will bring profound changes to the globe as a whole — but the particular types and scales of these impacts will be profoundly affected by local conditions. Climate change won't just have general global effects — over time it will create specific impacts within your very own neighborhood. Because the changes people will face vary from place to place, local governments (city or regional) will be best equipped to address these problems — one-size-fits-all solutions won't work. National governments that signed and ratified the UN Framework Convention on Climate Change (see Chapter 11) agreed to undertake adaptation planning. Several countries, such as the Dominican Republic, Cuba, and others in the Caribbean, are also undertaking regional planning. But fundamentally, although climate change is happening globally, people need to react locally.

Miami-Dade County in Florida is a leading example of what local governments can do. Not only is the county committed to reducing emissions, it's trying to make sure that people are prepared for the climate changes predicted, including sea level rise, salt-contaminated drinking water, and erosion. Miami-Dade has formed a Climate Change Advisory Task Force to make recommendations. As one of a handful of local governments in the United States already worrying about this problem, Miami-Dade is serving as a pilot project for Climate Resilient Communities, led by the International Council for Local Environmental Initiatives (ICLEI). (See the sidebar "Local leadership," in this chapter, for more about the ICLEI.)

One tragic example of a failure to adapt is New Orleans. In light of global warming, the city council had just realized that they needed to take action to protect the city from extreme weather events. In the summer of 2005, the Mayor spoke out as a signatory to the Mayors' Climate Protection Agreement, which endorsed the Kyoto Protocol. (See the section, "Success Stories," later in this chapter, for more about this agreement.) He pointed out that New Orleans was one of the most vulnerable cities in the United States to severe weather events, and would be even more vulnerable given the trends of global warming. The city realized that they needed to repair the dikes and levees, and beef up emergency planning, but Katrina hit before they could make the changes.

Local leadership

The Local Governments for Sustainability (formerly the International Council for Local Environmental Initiatives) — known as ICLEI for short — started as a partnership between 200 city and town governments from around the world in 1990. It's grown into a larger partnership of over 700 governments that share a strong commitment to sustainable development. Their projects have a large range of objectives, including the following:

- **Cities for Climate Protection:** ICLEI's flagship campaign puts policies and practices in place to reduce the city or town's greenhouse gas emissions and improve the city or town's quality of life.
- **Sustainable Communities and Cities:** This program's goal is to help cities foster justice, security, resilience, viable economies, and healthy environments.
- **Water Campaign:** This program measures how much water a city or community is using, develops a plan that provides targets for efficient water use, works to meet those goals, and continually measures and improves those goals. Water use and infrastructure require a lot of energy. By conserving water, you can conserve energy and reduce greenhouse gas emissions.
- **Biodiversity Initiative:** This initiative increases the local government's role in conserving the diversity of plants, animals, and entire ecosystems within or around a city or community. In the same way that you try to keep your immune system healthy so that your body can deal with and recover from a fever, this biodiversity project keeps ecosystems healthy so they can better deal with climate change.

Here are several examples of actions that regional governments can take to help their citizens adapt:

- **Diversifying crops:** Encouraging farmers to grow a wide range of crops helps governments protect both the food supply and farmers' incomes by ensuring that those farmers don't put all their eggs in one basket, so to speak. For example, if a pest that became rampant because of high temperatures causes the barley crop to fail, the farmers will still have the food and income from the corn and wheat crops that the pest didn't affect.
- **Expanding green space and tree cover in cities:** This measure absorbs runoff from large rainfall and reduces local temperatures, lowers demand for air conditioning, and stores carbon dioxide within the growing trees and associated soil. (Flip back to Chapter 2 for carbon cycle fundamentals.)
- **Improving floodwater management systems:** This involves taking extra measures to deal with more extreme and frequent flooding — such as making sure the street sewer system is working and maybe adding more street draining systems. Less asphalt and more green space in cities

slows rapid runoff to storm sewers, reducing erosion and downstream flooding. Governments might also want to rethink and rebuild water treatment plants; the water systems were built for a different climate.

- **Improving or installing storm-warning systems:** This involves preparing for more intense and more frequent storms — for example, by establishing a way to notify everyone when a big storm is coming.
- **Improving water management and watershed planning:** Cities need to plan for more extreme droughts and more intense river runoff, and ensure that they'll have drinkable water available for their citizens.
- **Increasing disaster relief funds:** Governments need to put aside more money for cleaning up after events such as hurricanes and flooding.
- **Securing levies against flooding:** Ensuring that the physical barriers protecting the city or town from flooding are strong and always being upgraded in advance of need enables cities and towns to continue to provide effective protection, even as the risk and intensity of flooding increases over the course of the 21st century.
- **Setting building restrictions on coastlines:** Governments can prevent the development of buildings too close to the water's edge to avoid loss of property due to erosion, which will be accelerated due to rising sea levels that will come with climate change.
- **Shifting tourism and recreation opportunities:** Tourism draws can be adapted to the changing climate and what the climate might be like in ten years. Some ski hills, for instance, may no longer be in operation because rising temperatures will rob them of reliable winter snow cover.
- **Subsidizing farmers who may need to relocate:** Governments may need to support farmers who have to move unexpectedly because their crops are no longer growing in the changed climate.

The cost and effectiveness of planning for adaptation are uncertain, but governments can gain insight from other groups' experiences. Australia, for example, has been emphasizing water management. Inuit communities in northern Canada are hunting differently to adjust to a changing climate, but are maintaining their culture. The Netherlands is building even more coastal barriers to protect against rising sea levels. Ski resorts across the United States, Canada, Europe, and Australia are relying more on artificial snow to keep the slopes open.

Adaptation isn't a response to just one particular storm or one melting glacier. It's a way of thinking, a kind of climate mindfulness that governments need to integrate into the plans they're already making. Australia already had a water management plan before climate change came along; climate change just added an extra angle — and urgency.

Cleaning up transportation

Although governments don't control everything that has to do with transportation, they still have a fair amount of influence over it. And when it comes to cutting greenhouse gas emissions, transportation is one area in particular in which the industrialized nations can make dramatic changes — largely because they use some of the most inefficient forms of transportation on the face of the Earth!

Different levels of government control different pieces of transportation. City governments, for example, control the public transit and city vehicles, which include public-service vehicles, such as park vehicles, snow plows, and garbage trucks. National governments can decide that a vibrant national rail system matters — moving people quickly and reducing greenhouse gases. National governments can also control what kind of cars are sold in the region or in the country, where and whether highways or rail lines are built or upgraded, and how many lanes of traffic are dedicated to multiple-passenger cars on the highway.

Bringing back the bike

Many cities are recognizing that bicycles can provide an environmentally friendly alternative to cars on the road. Municipal initiatives to encourage cycling range from improving bike paths to routing cars away from downtown areas. Many North Americans don't view the bike as a year-round transportation solution, but the Dutch and the Danes see things differently; a large percentage of city dwellers cycle contentedly throughout the cold and wet winters of northern Europe. Getting the right kind of bike and having dedicated bike lanes helps to extend the length of the bike-riding season in just about any city.

One of the world's most bike-friendly cities, thanks to its local government, is Amsterdam. There, you find separate bicycle lanes for each direction, with their own traffic lights that sync with those for cars. Spots where you can lock up your bike abound. City buses are equipped with bike racks, which passengers can use to stow their bikes when they board the bus. This sort of *intermodal transport* — switching between different ways to get around — is critical because it gives commuters convenient and flexible options, and makes it easy for people to not depend on their cars. (Pedal over to Chapter 17 for more about the benefits of bicycles.)

Investing in public transportation

Simply by providing adequate mass transit, municipal governments play a big role in reducing greenhouse gases. Driving in stop-and-go city traffic produces the majority of car emissions. The IPCC says that in a high-density city (which is ideal for public transit), 10 percent more buses on the road, filled with people who would otherwise drive their car, could reduce car emissions by 9 percent. But governments can do more.

The next French revolution: Velib!

When Paris celebrated the anniversary of the French Revolution, the city launched a new revolution (no guillotines involved) — a bike revolution! July 15, 2007, saw the opening of a city-run bicycle rental service planned to grow ultimately to 1,400 rental stations offering more than 20,500 bikes for hire. Everything is computerized — you just swipe your card, and away you ride. Plus, the first half hour is free. Many universities and cities around the world are taking this same initiative — making it easier for people to pick up a bike anywhere and ride. (*Velib* is a French play on words for "bicycle freedom.")

Many cities now have energy-efficient public transit buses. Your municipal government can run its public transportation fleets on alternative fuels or blends. In Halifax, Nova Scotia, whenever a bus goes by, you smell fish and chips coming from the tailpipe because its city buses run on recycled frying oil. Cities can also use gas-electric hybrids and good old electric trolley buses. (See Chapter 13 for more on alternative fuels.)

To get people to ride their energy-efficient public transportation, municipal governments can encourage riders in a number of ways. In London, youth under the age of 18 travel for free on major bus routes. Some countries, such as Canada, offer tax rebates to people who buy transit passes. Other regions create dedicated lanes on their roads to ensure a swift trip for public transit users. If it's faster to take the bus, why drive?

Buses aren't the only form of public transportation. Trains are already one of the most efficient modes of transportation, and they have the potential to become 40 percent more efficient by using new technology, according to the IPCC. (We take a closer look at public transportation in Chapter 17.)

Greening gas

Fossil fuel, which powers just about all vehicles, is the culprit for much of the world's greenhouse gas emissions. (Refer to Chapter 4 for more about fossil fuels.) Although governments can't force everyone into hybrid or electric cars, they can stipulate just how much fossil fuel goes into gas and how efficient cars must be at using that gas.

Governments can regulate fuel content, mandating that all gas sold at the pump has to contain a certain minimum percentage of an alternative fuel. For example, the other day, Zoë filled up at a local gas station to find that the regular fuel was now 25 percent ethanol, which meant her vehicle emissions were cut before she even restarted the engine. (Most ethanol produced in North America does have some major problems, which we talk about in Chapter 13, so buyer beware.) This type of regulation also creates a market for alternative fuels such as ethanol and bio-diesel. (See Chapter 13 for more on alternative fuels.)

The International Energy Agency suggests that governments need to use subsidies and trade policies to increase the production of low-emissions fuels and technology. If your government adopts this approach, some of your taxes that are currently going towards oil and coal products would go to alternative fuel technology. We discuss how governments can fund research in the section "Putting programs into place," earlier in this chapter.

Dealing with personal vehicles

Many people have a love affair with their cars, we know. But although we hate to break up a beautiful relationship between someone and their SUV, cars are big contributors to greenhouse gas emissions from transportation. Governments could definitely help nudge the auto industry and people toward making smarter transportation decisions.

Although most governments already regulate fuel economy, they could put the pedal to the metal and step up those requirements, demanding that a tank of gas take the car farther. In the 1970s, in the aftermath of the first oil shock, the United States first introduced regulations for cars, requiring all domestically produced cars to exceed an average 27.5 miles per gallon.

The IPCC says that national governments could potentially cut their country's car-related greenhouse gases in half by 2030 just by requiring automobile manufacturers to produce more energy-efficient cars. Check out Chapter 17 for more on car fuel economy.

Municipally, cities could make parking spaces smaller. If your vehicle is too big, you pay for two parking spots, rather than one! Alternately, cities could reduce the number of parking spaces they require in new developments — particularly if they're located close to transit services, increasing the value of parking spaces, and the convenience of not needing to find — and pay for — a parking spot. These and other measures could encourage people to drive smaller, more efficient cars. The City of Portland, Oregon, has gone one better. It has virtually eliminated downtown parking spaces!

More cities could pass anti-idling bylaws, ticketing people caught leaving their engines running unnecessarily.

Redefining long-term investments

The government makes investments with its money just like you can. But the government doesn't put its funds into stocks and bonds. Rather, governments select industries or economic sectors that they feel benefit their citizens one way or another, perhaps in a perfectly straightforward fashion (for example, by providing jobs) or in a less immediate, more long-range way (by encouraging farmers to stay on the land, for instance).

Most governments invest in (or *subsidize*) the energy sector. Typically, this investment means that they provide funds to oil companies, but governments could use subsidies to send the energy sector in very different directions. Governments could financially support companies that develop renewable sources of energy, rather than supporting those searching for fossil fuels such as oil. According to the IPCC, reducing fossil fuel subsidies is an effective way to wean the industrial world off its oil addiction. The IPCC warns that oil will remain civilization's primary fuel source as long as it remains subsidized by governments. The IPCC suggests reducing these subsidies (along with establishing taxes or carbon charges on fossil fuels and providing subsidies for renewable energy producers) as measures towards shifting to renewable energy supplies.

According to the European Renewable Energy Council (EREC), taxpayers worldwide currently subsidize oil, coal, gas, and nuclear energy to the tune of $250 to $300 billion annually, with some of these subsidies guaranteed for up to two decades. EREC notes that the imbalance between the subsidies that these fuels get and what is spent on renewable energy sources will hamper the development of sustainable fuels.

If governments want to reduce greenhouse gas emissions, they need to shift subsidies toward low-emission renewable energy. This shift would be more than an investment for the current citizens of the country; it would be an investment for future generations.

Governments are already moving in this direction. The IPCC reports that coal subsidies have dropped in the last decade around the world. By shifting such subsidies to renewable energy sources, governments can help those sources develop — at no additional cost to the taxpayer. Many governments are already beginning to invest in renewable energy, including the following:

- **Australia:** In 2000, the Aussie government developed a $303 million program that would cover half of the start-up cost of renewable and off-the-grid power generation. It started as a way to reduce dependence on diesel fuel and has since grown to reduce dependence on all fossil fuels.
- **Finland:** The national program called Energy Aid, specifically for companies and corporations, has been in place since 1999. It covers 25 to 40 percent of renewable energy and energy conservation projects.
- **Germany:** Through the Renewable Energy Sources Act, the government puts the end-consumer cost of renewable energy on par with other types of energy by sharing the cost with the energy company.

- **United Kingdom:** In 2002, the government initiated a large-scale solar demonstration project, paying $63.6 million to install photovoltaic solar systems. These funds covered 55 percent of the costs for public projects and 40 percent for large companies. Building on the success of that project, the government committed $172 million dollars in 2006 toward energy-efficient buildings that generate their own power. Both businesses and public places, such as schools and community centers, can get government funding to install solar panels, heat pumps, or other major energy-saving technologies that directly power their buildings.

Solar, wind, tidal, and geothermal energy (see Chapter 13) used to be futuristic ideas, but countries around the world are using them today. In fact, in 2005, these forms of energy produced 2.2 percent of the world's electricity — 18.2 percent when hydroelectricity (mostly old-fashioned large hydro) is added in. This statistic could rise to between 30 and 35 percent by 2030 (including hydroelectricity) if the cost of renewable energy continues to drop — and government investment can go a long way toward ensuring that it does.

Laying Down the Law

Regulations and taxes: hardly the way to win new friends. But as controversial and unpopular as they can be, they're vital tools for governments that want to control greenhouse gas emissions. Governments can regulate everything from the amount of energy your refrigerator can use, to how much insulation your house needs, to the amount of gas it takes to move your car a mile (or a kilometer). Taxes can be a powerful disincentive to individuals and companies; stop polluting, or you have to pay!

What level of government regulates what activity differs from country to country — and sector to sector. For example, in Australia, Canada, the United Kingdom, and the United States, the federal government has the power to set fuel efficiency standards, whereas state, provincial, or territorial governments have control of building codes.

Improving building regulations

The way houses and larger buildings are constructed is as much a product of policy as it is engineering. Building regulations apply to a building's structural elements, the materials used, their proper installation, and so on — but regulations can also control home energy use. By stipulating how the house is insulated and what sort of doors and windows to use, building regulations can determine whether your new home will be an energy guzzler or an energy miser over its lifetime. (We look at how buildings can be environmentally friendly in Chapter 14.)

The honor system: Self-regulation

One very basic form of regulation is *self-regulation,* in which the affected parties police themselves. As you can probably guess, politicians prefer self-regulation to putting emission limits in place. Having someone volunteer to do something is much easier than having to tell them to do it.

The major benefit of this type of regulation is simplicity. First, industries promise to reduce their greenhouse gas emissions below a self-determined baseline before the reductions. Then, governments check to see whether they're being compliant. If they are, great. If not, well, better luck next time — so, self-regulation doesn't guarantee greenhouse gas emissions reductions. Compliance is voluntary, and industries can easily set their reduction goal at zero.

Although successes are possible, self-regulation usually doesn't work, and it hasn't been proven to work at all on a larger scale. China is attempting to make it work by partnering with its 1,000 biggest industries, which draw on a third of the country's energy. The industries will measure, report on, and reduce their energy use through conservation and energy efficiency. The government will support the industries by promoting energy-saving initiatives, monitoring the progress, and helping to develop the training and implementation of the projects. Although it's a step in the right direction, the program doesn't have any specific targets.

When it comes to battling climate change, the IPCC reports that how people build, insulate, and heat their buildings has more potential to reduce greenhouse gas emissions than changes in energy, industry, or agriculture. With the right regulations in place, building-related emissions could be cut 30 percent by 2030 — without affecting the profitability of the construction industry.

In addition to improving standard building features, governments could demand that new buildings include certain elements that would make those buildings green (not to mention much cooler — or warmer, depending on the season). They could mandate that all new houses and other buildings include solar panels to help heat and cool them, or that contractors build *smart homes* — buildings equipped with automated systems that control heat and lights to conserve energy.

Regulating energy use

National and local governments hold great regulatory power over energy use. Greenhouse gas emission reduction targets are set internationally, but national or even more local levels of government have to regulate people's and companies' behavior to meet these targets.

Here are a few ways that governments can encourage more responsible energy consumption through regulation:

- **Clear labeling:** Governments can require producers to label all appliances and mark how much energy they use, and — most importantly — governments can decide what level of energy use qualifies as efficient. A producer would have to meet that requirement before selling their products as efficient.
- **Measuring usage:** Measuring is the first step to setting a target for reduction. California has installed *smart meters* in homes, which allow electricity consumers to see on an hourly basis how they're doing in their conservation efforts. Studies have proven that when people can see how much energy they're using, they cut back on their consumption. Watching the costs add up helps people remember to avoid *peak demand* periods (when more consumers are pulling power at the same time) for some energy-draining activities.
- **Setting quotas:** Governments can stipulate that major users of energy purchase a certain percentage of their power from renewable sources. Australia, for example, requires major buyers of electricity — such as industries — to buy a portion of their energy from wind farms and solar-power producers. If a user doesn't meet the quota, it must pay a fine.
- **Setting targets:** Governments can encourage the development of renewable energy sources by setting targets for the entire country. Industrialized member countries of the Kyoto Protocol, for example, each have a target of their own, as well as a collective global target of 5.5 percent below 1990 levels by 2012. Governments can also require business and industry to set a target to lower emissions.

Taxing the polluters

"There is no such thing as a good tax," Winston Churchill once said. We respectfully disagree. The most significant government power is taxation. Government uses taxes not only to raise revenue for the state, but also to discourage certain forms of behavior — think of the taxes on cigarettes and alcohol. Governments in some countries apply a similar tax policy to greenhouse gas emissions.

Governments could encourage people to stop producing greenhouse gas emissions by placing a fee on those emissions. The economic impact of such taxes has been minimal, even positive, as long as this fee is offset by reducing taxes on other things, such as income, jobs, and profit. This tax reassignment is called *tax shifting*. It's not about more taxes; it's about different taxes.

Most economists agree that this sort of tax, commonly called a *carbon tax,* is the most sensible and cost-effective approach to reducing greenhouse gases. No less an economic expert than the former head of the U.S. Federal Reserve, Paul Volcker noted that the argument that says taxes on oil or carbon emissions would ruin an economy is "fundamentally false. First of all, I don't think it is going to have that much of an impact on the economy overall. Second of all, if you don't do it, you can be sure that the economy will go down the drain in the next 30 years."

So far, carbon taxes are far more common in Europe than in North America. Norway has reported great success in using taxes to reduce both carbon dioxide emissions and HFC/PFC (the most powerful of the greenhouse gases) emissions. The most competitive economies in the EU have carbon taxes. Today, only three jurisdictions in North America have some form of carbon tax: Hawaii, Quebec, and British Columbia. Hawaii adds a surcharge to every barrel of oil shipped into the island state. The provinces of Quebec and British Columbia have a more universal carbon tax, and British Columbia's approach includes money back to the taxpayer. Now endorsed by the *Los Angeles Times* and *New York Times,* carbon taxes are likely to make their way into the U.S., as well.

TECHNICAL STUFF

The "carbon" in carbon tax can refer either to carbon dioxide emissions alone or to *carbon dioxide equivalents* — meaning all greenhouse gas emissions measured as a factor of carbon dioxide. The carbon market in Europe, for example, considers only carbon dioxide. New Zealand uses a carbon market that includes carbon dioxide equivalents. Both systems work, but those that use carbon dioxide equivalents can potentially have a greater effect because they include all greenhouse gases.

CONTROVERSY

The IPCC warns of a couple of problems associated with taxing greenhouse gas emissions: The taxes can't ensure actual emission reductions (polluters may be perfectly willing to pay to keep on polluting), and the taxes are tough to put in place from a political point of view because few things are as unwelcome to voters as taxes. On the other hand, carbon taxes can be implemented quickly, and are simple to administer, both of which make them attractive to governments gutsy enough to introduce them. All things considered though, taxes offer an effective emissions-reducing tool.

London's levy on cars entering the downtown area has been a very successful measure to reduce greenhouse gases, improve air quality, and relieve downtown congestion. The initial protests let up because Londoners agreed that the city was vastly more livable with fewer cars clogging the heart of the city.

Success Stories

Throughout this chapter, we talk about what governments can or should do to help reduce greenhouse gases. Don't get us wrong, however — governments all over the world, at every level, are already doing leading-edge work, moving toward low-carbon technologies and ways of life.

Cities and towns

You hear "Think globally, act locally" a lot these days — you need a big-picture perspective, but you have to make changes on a small scale, within your own city or town. Cities in industrialized countries can really reduce local greenhouse gas emissions, as many cities have already discovered.

Cities often act as a team to support each other in climate initiatives. In the United States, the Mayors' Climate Protection Agreement (the brainchild of Seattle's mayor, Greg Nickels) has a target similar to the Kyoto Protocol's — to reduce city emissions by 7 percent below 1990 levels by 2012 (for more about the Kyoto Protocol, see Chapter 11). More than 680 city mayors, representing over 74 million citizens, have signed up so far. These mayors are making progress, even if the national and state governments in the United States aren't committed to Kyoto targets.

Check out Table 10-1 for a list of cities that are leading the way to climate change solutions as members of the Cities for Climate Protection project, organized by the ICLEI.

Because so many factors contribute to greenhouse gases, many cities are concentrating their emission-reduction efforts on one aspect at a time. Rayong in Thailand installed a biogas facility to deal with the city's waste and to create an alternative fuel source. Mareeba Shire Council in Australia is using heat-reflective paint on the roofs of major city buildings to keep them cool, thus reducing the energy used for air conditioning.

Transportation is another area in which cities are cleaning up. Green Fleet, a project of the Cities for Climate Protection initiative, reduces emissions from cities' fleets of cars and trucks by ensuring that they're driven less, eliminated when possible, and that newly purchased vehicles are lower-emission. Hyderabad in India is improving traffic flow to lower the amount of time cars spend on the road and therefore reduce emissions. In Curitiba, Brazil, what began with the determination of its mayor to improve the bus system so that it would work better led to a phenomenal urban redesign that is centered around public transportation that is widely used, largely eliminating the need for cars. This initiative also made the city a more desirable place to live, increasing its tax base and quality of life. (For more about Curitiba, see Chapter 12.)

Table 10-1 Low-Carbon Cities

City	Population	Successes	Targets
Austin, Texas, United States	692,000	Cutting energy emissions 8% in five years; Improving energy efficiency by 7%; Saving $200 million through energy conservation	Getting 20% of energy from renewable sources by 2020; Improving energy efficiency by 15% (of 1992 levels) by 2020
Berlin, Germany	3,500,000	Cutting GHG emissions 14% (of 1990 levels); Saving $2.7 million a year	Cutting GHG emissions 25% (of 1990 levels) by 2010
Cape Town, South Africa	3,103,000*	Supplying clean and reliable energy to low-income households; Initiating top-of-the-line community housing project as part of the Clean Development Mechanism of the Kyoto Protocol	Cutting GHG emissions 10% by 2010; Getting 80% of energy from renewable and natural gas sources by 2050; Installing high-efficiency light bulbs in 90% of homes by 2010
London, United Kingdom	7,172,036*	Cutting CO_2 emissions 7% in ten years; Reducing transportation CO_2 emissions 19% in one year; Making $343 million in one year through fees	Cutting CO_2 emissions 20% (of 1990 levels) by 2020; Cutting CO_2 emissions 60% (of 2000 levels) by 2050; Building a development in each suburb that creates no overall carbon dioxide emissions by 2010
Melbourne, Australia	3,417,200*	Cutting GHG emissions 16% in six years; Getting 23% of energy for corporations from renewable sources	Allowing zero net emissions from corporations by 2020; Getting 50% of energy for corporations from renewable energy sources by 2010
Toronto, Canada	2,481,494*	Cutting GHG emissions 2% in eight years; Cutting corporate GHG emissions 42% in eight years; Making $16–$25 million from capturing landfill methane gas	Cutting GHG emissions 20% (of 1990 levels); Getting 25% of energy for corporations from renewable sources

Source: The Climate Group, Municipal Solutions. `http://theclimategroup.org`.

**Source: Encyclopedia Britannica Online, 2007.* `www.britannica.com`

The power of town hall

The Alliance of World Mayors and Municipal Leaders released a declaration in 2005 at the UN Climate Change Conference in Montreal, announcing long-term goals that are among the most ambitious in the world: They want to cut their cities' greenhouse gas emissions 30 percent below 1990 levels by 2020 and 80 percent by 2050. Their short-term target is a 20-percent reduction below 1990 levels by 2010.

Building on that declaration, concerned mayors and municipalities created the Cities for Climate Protection program. This program works with over 800 cities that account for approximately 15 percent of global human-caused carbon dioxide emissions. The project is lead by the International Council for Local Environmental Initiatives (ICLEI, also known as Local Governments for Sustainability). The initiative involves a five-step commitment:

- Keeping track of and predicting greenhouse gas emission levels
- Setting a greenhouse gas emissions reduction target
- Making a plan for how to reduce those greenhouse gas emissions
- Taking action on the plan and actually reducing the greenhouse gas emissions
- Keeping track of reductions and reporting on progress

You can get a taste for what some of these 800 cities are doing to reduce greenhouse gas emissions by checking out Table 10-1.

States, provinces, and territories

Governments at the state, provincial, or territorial level have a very important role in fighting global warming. Regional governments that have had success tackling their greenhouse gas emissions have emphasized renewable energy standards, establishing programs to encourage energy efficiency and creating cap-and-trade programs for industry.

Successful solos

Going solo takes a lot of guts — whether it's presenting to a group of people, performing on a stage, or doing something as minor as saving the world from climate change. Regional governments, such as those in the following list, are world leaders in greenhouse gas reductions, often without direction from their federal governments or any other regional governments:

- **Flanders, Belgium:** In this region, every electricity producer has to buy a certain amount of renewable energy. This policy started with a quota at 0.8 percent of its electricity in 2002, and it plans to build to 6 percent in 2010. The regional government fines producers that fail to meet this quota; the fees are fed directly back into Flanders' Renewable Energy Fund.

- **Victoria, Australia:** This state has its own $92-million Greenhouse Strategy. This money went directly into climate change projects between 2002 and 2004. The overarching goals were to build public awareness of climate change, reduce overall greenhouse gas emissions in Victoria, research climate impacts and adaptation strategies for the region, and look into opportunities for industry and business to reduce emissions. The result? Victoria's industries are cutting emissions by 1.23 million metric tons a year, a regional building standard is in place that sets the bar for energy-efficient homes, and the region contains some of the largest wind farms in the country. Next, the state plans to develop one of the largest wind farms in the world!
- **Walloon, Belgium:** Since 2003, the government of this region has required every electricity plant to get at least 3 percent of its energy from renewable sources, and it plans to build this percentage to 12 percent by 2012. If a plant doesn't follow the requirement, it's fined. That cash goes straight to Walloon's Energy Fund. Renewable energy producers can apply for support of about $100 per megawatt hour, which the Energy Fund finances.

Powerful partnerships

Like cities, some regional governments have found strength in partnerships. The U.S. state of California and the Brazilian state of Sao Paulo decided to work together to fight greenhouse gas emissions because, as their climate agreement notes, they're in very similar situations. Both states

- Are major contributors to their respective national economies and have the largest state populations in their countries (both boast over 35 million people)
- Have the highest energy use in their countries
- Suffer from major air pollution, which makes them interested in energy-efficiency measures for their regions
- Are regarded as leaders in developing and implementing programs to lower car emissions

California and Sao Paulo's agreement has three major commitments:

- Reduce the effects of air pollution
- Adapt to global warming by reducing the dangers their economies, people, and natural environments face
- Put policies in place that are stricter and cleaner than current measures

The emission terminator: California

California's announced target of an 80-percent reduction in greenhouse gases against 1990 levels by 2050 is ambitious (California is the 12th-largest emitter of greenhouse gases in the world), but achievable. Governor Arnold Schwarzenegger plans to wrestle greenhouse gases to the ground (and it won't be thanks to rippling muscles or Hollywood special effects).

The California Global Warming Solutions Act of 2006 requires that the California Air Resources Board (CARB)

- Establish a statewide greenhouse gas emissions cap at 1990 emission levels by 2020, based on 1990 emissions.
- Adopt mandatory reporting rules that control significant sources of greenhouse gases by January 1, 2009.
- Develop a plan by January 1, 2009, indicating how regulation, market mechanisms, and other actions will reduce emissions from significant greenhouse gas sources.
- Adopt regulations by January 1, 2011, to achieve the maximum technologically feasible and cost-effective reductions in greenhouse gas, including provisions for using both market mechanisms and alternative compliance mechanisms.
- Convene an Environmental Justice Advisory Committee and an Economic and Technology Advancement Advisory Committee to advise CARB.
- Notify the public and offer them an opportunity to comment on all CARB actions.

The two states are working together on a number of projects. For example, California is working with Sao Paolo to implement a project to clean the air, using the same framework that California did with its Federal Clean Air Act. At the same time, Sao Paolo is working with California's planners to help replicate Brazil's successful Bus Rapid Transit in California. A lot of the partnership is based on sharing information — whether about ethanol, substituting diesel with natural gas, conserving state forests, or generating electricity from biomass.

The Regional Greenhouse Gas Initiative (RGGI) is another local government initiative aimed at reducing greenhouse gases. Made up of ten northeastern and mid-Atlantic U.S. states, from Maine down to Maryland, their regional target is a 10-percent reduction in carbon dioxide emissions from power plants by 2018. They plan to do this through a cap-and-trade system, which we discuss in the section "Planning for emission trading," earlier in this chapter.

The action plan was consistent with climate goals that the Council of New England Governors and Eastern Canadian Premiers negotiated for a larger region. The RGGI was the first plan to call for hard caps to reduce greenhouse gas emissions in the U.S. To reduce emissions, the New

England–Eastern Canada accord pushed for more energy-efficient lighting and public awareness programs, and it's acting more quickly on greenhouse gas emission reductions than the federal government of either nation.

The Western Climate Initiative (which includes four Canadian provinces and seven U.S. states) works in a similar way. Their goal is to reduce greenhouse gas emissions to 15 percent below 2005 levels by 2020. Although this initiative isn't as strong as Kyoto or the reductions that California, Montana, Oregon, and Washington have already committed to, it nevertheless has three additional U.S. states (Arizona, New Mexico, and Utah) focusing on a reduced-carbon future. Its newest project is implementing a cap-and-trade system, which it plans to use with other carbon trading systems, such as that of the European Union and the Regional Greenhouse Gas Initiative.

Countries

Many countries have shown that nations can reduce energy consumption while growing economically. The key is to improve efficiency, doing more with less — or doing it differently. In North America, regional governments are taking the big steps. On a national level, European countries are leading the industrialized world in going green:

- **Denmark:** Significantly increased its use of wind power, with 18 percent of its electricity coming from wind as of 2005.
- **Germany:** Met its Kyoto targets and now boasts a burgeoning export industry in wind turbines (as does Denmark).
- **Switzerland:** A world environmental leader for a long time. In recent years, the Alpine republic has undertaken a number of initiatives:
  - Passing a federal law to cut methane emissions from waste
  - Legislating a cut in carbon dioxide emissions from energy to 10 percent below 1990 levels by 2010
  - Implementing an energy program, *SwissEnergy,* that works with cities to set energy-efficiency standards for buildings, appliances, and transportation, among other things (see the sidebar "SwissEnergy: How it works," in this chapter)
  - Providing *EcoDrive courses* that show people how to drive efficiently (thus burning less gas and creating fewer emissions)
  - Negotiating voluntary agreements for carbon dioxide reductions, which include tax exemptions

SwissEnergy: How it works

A program launched by Switzerland's federal government, SwissEnergy describes its goals as "promoting energy efficiency and the use of renewable energy" to cut human carbon dioxide emissions. The program's strength comes from the cooperative approach of four departments (transport, energy, environment, and communications), which work together with cities, industry, organizations, and businesses. SwissEnergy has cut Switzerland's consumption of fossil fuels by 7.9 percent and electricity consumption by 4.7 percent. The use of renewable energies has increased, too.

You can find more information about SwissEnergy at www.bfe.admin.ch. (The home page isn't in English, but you can access an English version by clicking the English link in the upper-right corner of the page.)

The Swiss government has also taken a sustainable development approach to transportation. Thanks to Swiss government programs, major freight trucks have cut their annual road mileage by 6 percent while the freight volumes have increased. The money saved from this mileage reduction (over $1 trillion per year) is put into railway infrastructure. And because of this boost in funding, train travel is predicted to grow at least 40 percent by the year 2030.

Switzerland also boasts numerous projects run by private organizations. The Climate Cent Fund, for instance, takes a one-cent levy on every liter (1/4 gallon) of gas that's sold, and literally every penny from this levy goes into a fund for projects that directly reduce emissions in Switzerland and around the world.

Governments can't solve the climate change challenge alone. Many other players — you, your neighborhood, businesses, and industries — all have to work together to reduce greenhouse gas emissions. (We talk about those other players in Part V.)

Chapter 11

Beyond Borders: Progress on a Global Level

In This Chapter

- Understanding the importance of global agreements
- Looking at the United Nations Framework Convention on Climate Change
- Putting the Convention into action by using the Kyoto Protocol
- Discovering the source of the science: The Intergovernmental Panel on Climate Change

Generally, when we discuss the effects that greenhouse gases have on the world's atmosphere, we prefer the term *climate change.* Global warming simply isn't an accurate description. But we do like one thing about this description: It reminds us that the problem we face isn't just local or national, it's global. Humanity is experiencing a global problem — and that problem requires a global solution.

The United Nations has a very important part to play in fighting climate change. It provides a forum for governments to work together and hammer out solutions to international problems. Global problem-solving is a long, slow process — and a thankless one much of the time. You may have read news articles about the countless international conferences on global warming and wondered what goes on at those meetings. Different nations bring competing agendas to the table; representatives from all nations must overcome language and cultural barriers; and national governments face pressures at home from business, organized labor, and opposition parties. The world's glaciers may be receding faster than international agreements can move forward.

And yet, despite all the impediments, the world's nations make progress. Even better, sometimes they enjoy huge successes, such as the international agreement to stop the destruction of the ozone layer (which we discuss in the sidebar "International agreements work: The Montreal Protocol," in this chapter). Our world today is a safer place because of global agreements.

Global agreements hold countries accountable for certain actions and give nations a set of rules enforced through United Nations international law.

The world's governments have been struggling with climate change for more than 20 years. The process has been painfully slow, and those governments still have a lot to do. But, right from the start, every country (well, almost every country) agreed that no one nation can solve the problem of climate change alone. In this chapter we go beyond the headlines you may read about these international agreements; we explore just why they're so important and what goes into making these agreements happen.

Why Global Agreements Are Important

Countries can do a lot to tackle global warming individually, as we discuss in Chapter 10. But the problem is far too great, and the solutions are far too complex, for countries to attempt to address climate change on their own. Each country is responsible for a portion of greenhouse gas emissions and has the ability to reduce global emissions anywhere from a fraction of a percentage up to 25 percent. But it is only with a collective effort that global emission can be reduced 50 to 80 percent. The world needs a global agreement to reduce greenhouse emissions and fight climate change because such an agreement can

- **Coordinate everyone's efforts.** A coordinated effort ensures that everyone's working toward the same goal, rather than charging off in all directions.
- **Create an accepted target.** Taking a big-picture approach allows nations to get a more accurate assessment of the actual impact of their emissions. Countries can then work together to determine targets for reduction that can make a difference. (Check out the section "Looking at the Kyoto Protocol," later in this chapter, to see how countries set their targets.)
- **Ensure a level playing field.** In this age of a globalized economy, countries' government officials can feel terribly insecure, worried that businesses might leave them for a more accommodating nation. Some government officials might worry that imposing emission regulations would scare off industries; if all countries commit to reducing emissions, those government officials don't have to worry as much. (We talk about regulating emissions in Chapter 10.)

- **Include the developing nations.** Because developing nations have limited financial resources, they can't undertake initiatives for sustainable development or adaptation independently. A global agreement allows for wealthier nations to help these countries prepare for the effects of climate change and industrialize in a way that won't contribute more greenhouse gases to the atmosphere. (Flip to Chapter 12 for more about the impact global warming has on developing nations.)
- **Increase the transfer of technology and experience.** When one country reduces emissions, it can share those best practices through international systems set up within the Kyoto Protocol.

The United Nations creates a global sense of understanding, and it provides a structure and venue for countries to work on issues together. It's the ideal arena for global agreements because it was designed for that very purpose. And it gets results. (See the sidebar "International agreements work: The Montreal Protocol" for an example of a successful global effort to reduce emissions.)

The United Nations Framework Convention on Climate Change

The year 1992 saw the biggest gathering of heads of government ever held — the Earth Summit in Rio de Janeiro, which drew together 170 countries and close to 20,000 people. Leaders rarely seen on the same stage, such as then-U.S. President George H. W. Bush and Cuba's then-president Fidel Castro, made first-of-their-kind commitments that continue to this day, on the topics of sustainable development, biological diversity, and — of course — climate change.

The summit's aim was to tackle climate change, so the leaders created the United Nations Framework Convention on Climate Change (UNFCCC). (In this case, a *convention* means a legally binding agreement — a statement of principles and objectives without specific numbers or measurable targets.) The Convention's main goal is to stop the build-up of greenhouse gases in the atmosphere before the level of those gases becomes dangerous. (The Convention actually includes the word "dangerous" to describe the level of greenhouse gases that must be avoided.)

Of course, *dangerous* is a relative term. If you were in Europe during the heat wave of 2003 that killed 30,000 people, or fleeing the British Columbia forest fires of 2004, you might decide that climate change has already made the world pretty dangerous. In order to make the term *dangerous* a little less relative, the Convention *parties* (the participating countries) rely on the Intergovernmental Panel on Climate Change (IPCC). (We talk more about the IPCC in the section "The World's Authority on Global Warming: The IPCC," later in this chapter.)

The UNFCCC lays out the groundwork for action on climate change by

- Committing all parties to a shared commitment to action (committing to a commitment is sort of like giving your sweetheart a promise ring pledging that you'll get engaged — you're promising to make a promise).
- Acknowledging that climate change is occurring and that human activities, such as burning fossil fuels and changing land use (such as deforestation), are the major sources of this change.
- Accepting that if the parties wait for 100-percent scientific certainty, the problem will be too advanced to fix. The Convention adopts the *precautionary principle* — that no party can use a lack of scientific certainty as an excuse for inaction.

Talking the global agreement talk

We talk a lot about conventions and protocols. And you've probably read about other conventions, too — the Geneva Convention, for example, on the rights and treatment of prisoners of war. In diplomatic terms, these words have very specific meanings — quite different from how people normally use them. In this context, a *convention* basically involves agreeing to a principle — in this case, fighting climate change — and setting objectives, but with no timelines or specific targets.

As soon as all the countries' government officials agree to the language of a convention, they normally sign it right away. Then, they take it back to their countries, where it must be ratified, or domestically approved. Ratification processes vary. In the United States, treaties must be subjected to a Senate vote where three quarters of the Senate must approve before the U.S. is bound by international law. In Canada and other parliamentary democracies, a treaty can be ratified by a simple Order in Council within Cabinet.

After a country ratifies a convention, that country becomes a party to that convention.

Conventions generally agree on a *ratification formula* — the number of countries that need to sign on before the convention is enacted (or, in U.N. language, *enters into force*) and becomes legally binding.

Later, parties can agree on a *protocol,* an agreement that's stricter and more detailed than the original convention. Like a convention, it's legally binding, but it contains actual deadlines and targets.

Establishing a game plan

The United Nations Framework on Climate Change Convention doesn't explicitly spell out how the parties involved should tackle climate change (that came later — we get into the Protocol that came from the UNFCCC in the section "Looking at the Kyoto Protocol," later in this chapter). Instead, the convention committed the parties to "aim towards" stabilizing the level of GHGs in the atmosphere. It set out two areas in which the parties need to act:

- **Adapting to climate change that can't be avoided:** The Convention acknowledges that, regardless of how much greenhouse gas levels drop, due to the increased greenhouse gas concentrations from human activity, the world can't avoid some effects from climate change. All countries are going to have to adapt to a changing climate regime. Some countries may need to plant drought-resistant crops; others may need to build higher levees and dikes in low-lying areas, and not rebuild on flood plains. (We talk about adaptation in Chapter 10.)
- **Reducing greenhouse gases:** The Convention calls this process *mitigating,* which means cutting emissions.

Dividing up the parties

The UNFCCC recognizes that most human-caused greenhouse gas emissions stem from industrialized countries, and, therefore, it states that those nations should take the lead to battle climate change.

Countries vary in terms of how much they add to the problem and how able they are to actually help fix it. The UNFCCC breaks nations into three groups based on this variety, and it has different expectations for each group. In convention-speak, this division is called "common but differentiated responsibilities and respective capabilities." The following list describes how the groups are distinguished from one another:

- **Annex 1 countries:** This group includes all industrialized (developed) countries such as Australia, Canada, the United Kingdom, and the United States. Annex 1 includes the following sub-groups:
  - **Economies in transition:** This group is made up of countries that are transitioning to a market economy — primarily former Soviet countries such as Hungary, Belarus, and Poland.

  - **Annex 2 countries:** The Annex 2 countries generally have the strongest economies — this grouping includes all Annex 1 countries, *except for* economies in transition. Parties of the UNFCCC expect these countries to contribute money, technology, and other resources to Non-Annex 1 countries.

- **Non-Annex 1 countries:** This group basically includes all industrializing (developing) countries, such as Brazil, China, and India. These poorer countries will have a much harder time adapting to the impacts of climate change than the wealthy industrialized countries in Annex 1 and Annex 2. They don't have the money for new technologies or programs, and they often have far more immediately pressing issues to deal with, such as war, famine, HIV/AIDS, or inadequate clean water. (We look into the challenges facing developing countries in Chapter 12.)

Historically, the greater a country's gross domestic product (GDP), the greater its volume of greenhouse gas emissions. Annex 1 (which includes Annex 2) countries face the most aggressive targets for emission reduction because industrialized countries produce the greatest amount of greenhouse gases.

Looking at the Kyoto Protocol

The United Nations Framework Convention on Climate Change (UNFCCC) showed the world it was serious about tackling climate change when it met in Kyoto in 1997. At this meeting, called the Conference of the Parties (CoP), countries agreed on the *Kyoto Protocol* — a commitment to decrease greenhouse gas emissions by a set amount and before a set deadline. CoPs still continue to this day, with CoP 14 being held in Poland in winter of 2008. United Nations negotiations are always tricky because they require *agreement by consensus* — every party needs to be on board with the decision. Negotiations can go into the wee hours of the morning, often without a break. Bleary-eyed negotiators stay glued to their microphones while the translators share the discussions in six official languages.

In Kyoto, Japan, one set of negotiations went on for 36 hours straight. But in the end, the bad food, rumpled clothes, and sleep deprivation were worth it. Building on the Convention, negotiators agreed to the Kyoto Protocol.

Setting targets

National governments within the United Nations began discussing lowering greenhouse gas emissions way back in 1990 at the first meeting to set up the UNFCCC. And ever since, the conventions and protocols under the UNFCCC have used 1990 emission levels as the base for setting reduction targets.

Going into the negotiations at Kyoto, the European Union demanded a 15-percent global reduction in greenhouse gases. The United States and Canada insisted on a much lower target. Ultimately, the parties settled on a global target for industrialized countries of 5.2 percent below 1990 global emission levels.

SITUATION CRITICAL

A global reduction of 5.2 percent below 1990 levels is actually a 24.2-percent reduction today, according to the World Bank, because of how much emissions have increased.

TECHNICAL STUFF

The Kyoto Protocol covers carbon dioxide (CO_2), methane (CH_4), nitrous oxide (N_2O), hydrofluorocarbons (HFCs), perfluorocarbons (PFCs), and sulfur hexafluoride (SF_6). (We cover these greenhouse gases in detail in Chapter 2.)

The countries that have binding targets for emission reductions are Annex 1 countries to the Convention — but not all Annex 1 countries are part of the Kyoto Protocol. (The United States, for example, did not ratify the Protocol.)

International agreements work: The Montreal Protocol

In 1987, the United Nations Convention on Ozone met in Montreal, Canada, to negotiate a protocol to reduce the release of chemicals that were depleting the *ozone layer* — the layer of upper stratosphere that protects the Earth from the sun's ultraviolet rays. Many industries used these ozone-depleting chemicals as refrigerants and the propellant in aerosols.

The Montreal Protocol, a globally ratified agreement within the United Nations, acknowledged that industrialized countries had created most of the problem, that they had the best technology to solve the problem, and that poorer countries still needed access to chemicals so that they could economically develop.

The agreement required industrialized countries to cut production and use of ozone depleters by 50 percent, and less-developed countries could increase their use by 10 percent. Ultimately, 191 countries — almost every country in the world — agreed to get rid of ozone depleters altogether within a specified time frame. But the industrialized countries had to take the first step. The Kyoto Protocol takes a very similar, if not identical, approach.

The United Nations recently reported that ozone-depleting chemicals have been drastically reduced thanks to the Montreal Protocol. Recent studies show that the hole in the ozone layer has stopped growing and the layer is on its way to recovery. At the rate the hole is shrinking, the layer should be healed by 2050.

The countries need to achieve the emission cuts outlined in the Kyoto Protocol between 2008 and 2012. The Kyoto Protocol refers to this five-year time frame as the *first commitment period*. Countries are negotiating another set of targets for the second commitment period. These negotiations are mandated to have a new phase of Kyoto negotiated by late 2009.

The European Union (EU) negotiated a collective goal of an 8-percent reduction below 1990 levels. Within the EU, countries with better economic potential and the capacity to cut emissions received higher reduction targets — such as Germany, which committed to a 21-percent cut, and the United Kingdom, which committed to a 12.5-percent cut (and they've already surpassed that target).

Ideally, the largest cuts in greenhouse gas emissions were applied to the biggest emitters, but this was not always the case: political will and popular support by country played a major role in assigning targets.

Table 11-1 shows the commitments that selected countries made to reducing their greenhouse gas emissions and how well those countries are meeting those targets.

Table 11-1 Countries' Kyoto Protocol Greenhouse Gas Level Targets

Country	*GHG Level Change in Relation to 1990 Level*	
	2012 Target	*2005 Achieved*
Canada	–6%	+25%
Germany	–21%	–18%
France	Stay at 1990 level	–2%
Sweden	+4%	–7%
Japan	–6%	+7%
Norway	+1%	+9%
United Kingdom	–12.5%	–15%

Source: United Nations Framework Convention on Climate Change. Total aggregate greenhouse gas emissions of individual Annex 1 Parties, 1990–2005 (excluding land use and forestry).

Adding flexibility

Before agreeing to the Protocol, some industrialized countries insisted on a number of compromises called *flexibility mechanisms* (or loopholes, according to many environmentalists). These mechanisms involve carbon credits. If a country lowers its emissions more than its target, it receives a carbon credit for the extra reduction. (It's similar to what happens if you pay more than you owe on your credit card bill; you get a credit that goes toward your next bill.) The country can decide to either apply this credit to the next commitment period (when it sets its second set of targets) or sell the credits to another country that's having trouble meeting its targets now.

Because of the flexibility mechanisms, countries can actually make money if they lower their emissions more than they said they would. Of course, the downside of this trading is that some countries would prefer to buy credits, instead of actually reducing their own emissions. The good news is that the mechanisms convinced more countries to commit to the Kyoto Protocol.

The Protocol set out three specific flexibility mechanisms for Annex 1 countries:

- **Clean Development Mechanism:** Annex 1 countries get credits for funding emission-reducing projects in developing countries. (See Chapter 12 for more.)
- **Emissions Trading:** Countries that exceed their reduction targets can sell their extra reductions (carbon credit) to other countries. (We talk about how governments can implement emissions trading nationally in Chapter 10.)
- **Joint Implementation:** Former Soviet Union countries such as Ukraine and the Czech Republic (the Economies in Transition group — refer to the section, "The United Nations Framework Convention on Climate Change," earlier in this chapter) can sell their credits to those Annex 1 countries that have reduction targets. Annex 1 countries can also get credits for funding projects that reduce GHG emissions in former Soviet countries.

Emissions trading is the most controversial flexibility mechanism because it benefits one party almost exclusively. The USSR broke into smaller countries in the years that followed its collapse in 1991 — industry declined, and so did the emissions. But because Russia's targets were based on its 1990 emissions, it has already reduced emissions below its Kyoto targets and can sell its carbon credits — though it's not actually reducing its current greenhouse gas emissions. Of course, Russia hasn't always been just Russia. It, along with other countries that were part of the Union of Soviet Socialist Republics (USSR) (such as Ukraine and Belarus), has its own benchmarks, which represent a proportion of the old Soviet total. These other former Soviet countries also have credits to sell. No countries have purchased carbon credits from Russia (or other former Soviet countries) yet; industrialized countries consider buying carbon credits from former Soviet countries an ineffective option because no actual greenhouse gas reductions are being made. The goal is to further reduce emissions, not pay for reductions from almost 20 years ago.

The Kyoto Protocol includes one other way participating countries can score carbon credits. Annex 1 countries can get credit for enhancing *carbon sinks* — those natural ecosystems (such as forests) that soak up carbon, keeping it out of the atmosphere. The Kyoto Protocol grants credits for undertaking *afforestation* (planting trees on previously treeless land) and *reforestation* (replanting on land that used to be forest). (Commercial logging can't receive these credits — you can't get credit for planting a forest after cutting one down!)

Opting out: The U.S.A.

The United States is a member to the UNFCCC, but it hasn't ratified the Kyoto Protocol.

The United States accounted for 25 percent of global carbon dioxide emissions in 1990 (slightly less by 2004 — at 20 percent — because of China's rising percentage). The U.S. signed the Protocol, meaning it set a target for itself, but the Kyoto Protocol was never forwarded from the president to the Senate for a vote, thus the U.S. never ratified the Protocol. In short, the federal government has not yet brought the Protocol into force nationally. The targets aren't binding until a country ratifies them. The U.S. withdrew from the Kyoto process altogether in 2001.

President George W. Bush said he wouldn't ratify Kyoto because it would hurt the U.S. economy. The U.S. also argued that the Kyoto Protocol is unfair to industrialized countries. Bush claimed that it puts major businesses and firms at a disadvantage because they're bound by tough regulations but their competitors in developing nations aren't.

Australia had signed the Protocol but didn't ratify it until late in the game, even though its targets actually allow it to increase emissions by 8 percent above 1990 levels. Opponents of the Kyoto Protocol argued that ratification could damage Australia's coal industry (the world's largest), severely harming the country's economy.

On November 24, 2007, Australia's voters tossed out Prime Minister John Howard in favor of Kevin Rudd's Labor Party. Rudd promised that, if elected, his government would ratify the Kyoto Protocol. On December 3, 2007, Australia became the latest nation to do so.

Another agreement: The Asia Pacific Partnership (APP)

The Asia Pacific Partnership on Clean Development and Climate (APP) is a coalition of Australia, Canada, China, India, Japan, the Republic of Korea, and the United States. The stated goal is to work together to develop and utilize new technologies that reduce air pollution, increase energy security, and sustain economic growth.

Controversy surrounds the APP because many people say it was created to undermine the Kyoto Protocol. The APP has no emission reduction plans or targets. The Bush administration of the United States spearheaded this initiative to involve the fastest-growing economies in the world, but the partnership sets out no emission reduction targets or deadlines, both of which are essential to combat climate change. Critics have attacked the partnership for putting effort into a second agreement that covers the same issues as the Kyoto Protocol — reducing greenhouse gas emissions.

Ratifying Kyoto

For the Kyoto Protocol to become officially active, 55 countries had to ratify it, and those countries had to be responsible for at least 55 percent of the world's greenhouse gas emissions in 1990. On February 16, 2005, the Kyoto Protocol became legally binding.

When we wrote this book, 177 countries had ratified the Kyoto Protocol. This list now includes virtually every industrialized and developing country (including Brazil, China, and India). The notable exception is the United States — see the sidebar "Opting out: The U.S.A.," in this chapter, for more about America's position on the Kyoto Protocol. Only the countries that have ratified the Protocol can discuss issues and vote at the annual Kyoto Protocol meetings.

Of course, when national governments and their leaders change, countries might have a change of heart on the Kyoto Protocol. To un-ratify, a country has to wait three years from the date that the Protocol entered into force, and then it has to submit a notification of withdrawal, which is confirmed one year from the date received. (If the country withdraws from the Convention, it automatically withdraws from the Protocol.)

Negotiations take a long time, but humanity doesn't have another way to solve complex global problems. Parties to the United Nations Framework Convention on Climate Change meet annually, and since the Kyoto Protocol came into force in 2005, Kyoto parties also meet annually. At the first meeting of parties to the Kyoto Protocol in 2005, the parties launched new negotiations to establish the next round of reductions to begin in 2013, when the first commitment period ends.

The World's Authority on Global Warming: The IPCC

Scientists around the world agree that global warming is happening due to human activity, but they don't all agree what its local and specific consequences will be. Thousands of studies are published every year that consider every conceivable aspect of climate change. To keep track of the top research, the United Nations formed the Intergovernmental Panel on Climate Change (IPCC) in 1988. The IPCC is a group that assesses current research and compiles it in reports that it issues every five years.

In November 2007, the Nobel Peace Prize was jointly awarded to the IPCC and Al Gore for their work on climate change.

Getting to know the IPCC

The IPCC is made up of 2,500 scientific expert reviewers from 130 countries. These reviewers (all volunteers) come from a wide range of scientific backgrounds and include 450 main authors and 800 assistant authors who work together to create the IPCC's assessment reports. The governments of countries who are member parties of the UNFCCC can select scientists to be on the IPCC, often based on nominations from organizations or individuals. From these selections, the IPCC Bureau chooses their reviewers, picking candidates primarily based on scientific qualifications. However, other factors also come into play — the Bureau attempts to ensure that different regions, genders, ages, and scientific disciplines are all represented.

The IPCC's assessment reports offer a big-picture look at the science about, causes of, impacts from, and solutions to climate change. The IPCC's scientists look at all the relevant *peer-reviewed* science in the world (meaning any scientific article reporting on recent research about climate change that a group of scientists, other than the writers themselves, approve) and integrate this material into their report. The IPCC authors decide what to include by consensus, ensuring a broad and conservative agreement about what gets in. When disagreements occur, the IPCC notes them within the report.

Reading the reports

IPCC reports inform global talks about climate change. The IPCC published the first assessment report in 1990. This report played a key role in encouraging the attendees of the Earth Summit in Rio de Janeiro to create the United Nations Framework Convention on Climate Change (UNFCCC) in 1992. (The section "The United Nations Framework Convention on Climate Change," earlier in this chapter, can tell you more about this Convention.) The second report, completed in 1995, helped spur the Kyoto Protocol of 1997. After the third report was issued in 2001, those awaiting a scientific authoritative statement couldn't deny the man-made causes of climate change. The fourth report, issued in 2007, had a greater level of scientific certainty and further confirmed humanity's role in climate change. That report began to give specifics about regional impacts.

The IPCC's reports are heavy reading (literally — each report has about 3,000 pages and weighs approximately the same as a newborn baby). With each report, the connection between climate change and human activity has become clearer — and the warnings about what will happen if people don't act have become stronger.

The UNFCCC relies on the IPCC reports in their decision making. The reports recommend greenhouse gas reduction targets, regional adaptation strategies to climate change, and technological opportunities that can help reduce climate change. The reports' top-end advice is trusted by UNFCCC parties.

The third report

The third IPCC assessment report was the first to state that there was only a 5-percent chance that climate change was entirely natural. To put it another way, the IPCC was 95-percent certain that human activities were intensifying natural climate change. This report showed that the greenhouse gases people were adding to the atmosphere helped cause major climate changes, more intense floods, droughts, storms, and water shortages. (Check out Part III for more about the effects of climate change.)

The third report was also the first to set a limit on greenhouse gas emissions that the world shouldn't exceed if it wants to avoid dangerous climate change impacts. (Flip to Chapter 3 for more information about working within this limit.)

The fourth report

The most recent report, the IPCC's fourth, was released in 2007. In this report, the scientists offer a detailed estimate of how global warming will progress over time if greenhouse gas levels go unchecked. The report maps out what effects people will feel in what parts of the world. It also considers what impact various greenhouse gas reductions can have on the progress of climate change. Additionally, the report emphasizes the serious impact global warming will have on developing nations, analyzing how efforts to reduce greenhouse gas emissions and adapt to climate change might affect those countries, and how sustainable development plays a role in responding to climate change in developing countries. (Most of the science that we discuss in this book comes directly from the fourth IPCC report.)

Reports for the rest of us!

The IPCC is constantly writing reports, in addition to the regular assessment reports, geared directly towards policymakers. They write these action-focused reports without using a bunch of science jargon, and anyone can access them.

These reports include

- **Carbon Dioxide Capture and Storage:** Explains carbon capture and storage, what the costs are, how it affects the environment and people's health, and barriers to implementing it.
- **Fourth Assessment Report: Climate Change 2007:** Covers the scientific facts about climate change, considers the impacts climate change will have on people and the environment, and suggests how we might reduce greenhouse gas emissions.
- **Safeguarding the Ozone Layer and the Global Climate System:** Explains how the ozone layer and climate are related, what gases cause holes in the ozone layer, and what kind of products give off these ozone-depleting gases.

To find the main reports, visit `www.ipcc.ch/ipccreports`. On the Web site, you can select whichever type of report you are looking for. Most every report written by the IPCC has a "Summary for Policy Makers" or "SPM" version of the report that is free for download.

Chapter 12

Developing in the Face of Global Warming

In This Chapter

- Understanding the challenges that climate change presents developing countries
- Seeing positive developments in China, Brazil, and India
- Working toward a sustainable future

The political climate surrounding global warming is incredibly unfair. Although the major contributors to global warming have historically been the richest, most industrialized nations, now that those nations are waking up to the dangers of global warming, they're trying to hold developing nations to environmental standards that they themselves did not face. Worse still, these developing nations face the same environmental challenges as other countries, but without the financial resources to prepare for them.

This chapter investigates the unique challenges that these countries face while they seek to develop their economies in the face of global warming. We look at some positive steps that China, Brazil, and India (three of the world's largest and most populous developing countries) are taking. Finally, we look at what initiatives developing countries can take to reduce their carbon emissions and adapt to a warmer world, and how industrialized nations can help pitch in.

Growing Concerns

The countries of the world are roughly divided into two categories:

- **Developed:** These countries, which are also known as *industrialized countries,* have a strong industrial base and a relatively high income per capita. The generally accepted grouping of industrialized countries includes Australia, Canada, Japan, New Zealand, the United States, and all countries within Europe. Many people would also consider countries such as Russia and Israel to be developed.

- **Developing:** These countries generally have a low per-capita income and little industry. In developing countries, life expectancy is lower than in industrialized countries, and an increasingly urbanized population is growing more rapidly than the population in industrialized nations. All developing nations are moving toward industrialization (that's why they're described as developing), but some are closer than others. Most countries in the world are considered developing.

Figure 12-1 highlights 50 of the world's least developed countries.

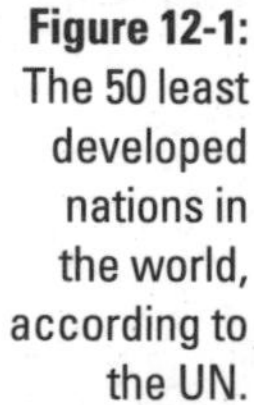

Figure 12-1: The 50 least developed nations in the world, according to the UN.

Based on The Least Developed Countries Report 2004 – Linking International Trade with Poverty Reduction. *UNCTAD.*

Some development academics argue that no country can ever truly be fully *developed* because progress has no real end. Others say that industrialization and a stable and strong economy signal that a country is developed. Another group thinks that development is defined by what educational and health services are available to the general population.

Developing countries face significant challenges while they try to build and improve their countries' economies. The route to wealth that all industrialized countries followed (such as developing big industry by using fossil fuels such as coal and oil) leads to climate change. While the wealth and industry in these huge developing countries grow, so too does their energy consumption. If these countries take their energy from the traditional sources that power industrialized countries, carbon emissions will skyrocket.

Consequently, the development of these countries is under great scrutiny. Some industrialized countries, such as the United States and Canada, have said that they won't commit to reducing their impact on climate change unless major developing countries, such as China, Brazil, and India, also commit. But the "Do as I say, not as I do" position from major industrialized nations doesn't sit too well with people in these countries. As Brazil's President Luiz Inácio Lula da Silva told *The New York Times,* "We don't accept the idea that the emerging nations are the ones who have to make sacrifices, because poverty itself is already a sacrifice."

The generally accepted idea, which is the core of the Montreal and Kyoto Protocols (which we discuss in Chapter 11), maintains that industrialized countries caused the problem and have more resources to tackle it, so they should take the first steps in fixing it. After they get a good start, then developing countries can join in. The technological innovations of industrialized countries can help make the transition to a low-carbon development path more feasible, and the developing nations can focus on reducing emissions when they're financially able to implement new technologies, including receiving financial and technological assistance from industrialized countries.

Promising Developments: China, Brazil, and India

China, Brazil, and India are three of the most heavily populated countries in the world: Their combined population is about 2.6 billion — close to 40 percent of the world's people. All three nations are generally considered developing countries. Yet each one is quickly moving toward industrialized status. Their development has led them down the same path that all other industrialized countries have traveled over the past 50 years — and industrialization, combined with growing populations, has led to massive increases in the amount of greenhouse gases they produce. China, Brazil, and India made up 22.2 percent of the world's carbon emissions in 2004, and their emission rates have since grown. The rapid pace of industrialization in developing countries gives plenty of cause for concern.

However, together, the United States, the United Kingdom, Canada, and Australia made up about 24.5 percent of the world's carbon dioxide emissions in 2004. This group of industrialized countries comprises just 6 percent of the world's population, which breaks down to an average of 21 metric tons of carbon dioxide per person per year. China, Brazil, and India collectively make up 40 percent of the world's population, which breaks down to an average of 7 metric tons of carbon dioxide per person per year.

When trying to figure out who's actually causing the carbon dioxide emissions of a country such as China, consider that the industrialized world imports many of the goods that country manufactures. One glance at the label on your shirt or the fine print under your breakfast bowl may tell you that it was made in China. The government of Sweden has publicly recognized that part of the industry emissions in China are growing simply to provide industrialized countries with things they want or need.

REMEMBER

Although China, Brazil, and India don't have to meet specific targets (yet), they (like all the countries involved in the UN Framework Convention on Climate Change and the Kyoto Protocol) have a general obligation to work to reduce emissions. Their governments understand that the climate crisis is real. (See Chapter 11 for more about the Kyoto Protocol.)

In the negotiations toward the second phase of the Kyoto Protocol (the first phase ends in 2012), these economic giants of the developing world are opening the door to targets. So far, the discussions fall short of national targets, but developing countries could set targets for individual groups of carbon emitters, including a target for reductions in the electricity-generation sector of China, for example. (See Chapter 11 for more on the Kyoto Protocol.)

Although China, Brazil, and India may not have any specific targets under the Kyoto Protocol, they ratified the Protocol, which means that they've agreed in principle to reducing greenhouse gas emissions. (The Kyoto Protocol requires industrialized countries to take the first steps.) If these countries act on their commitments, the results could be tremendous: In fact, the World Bank and the United Nations Environment Program (UNEP) report that China, Brazil, and India could potentially cut their collective emissions by 25 percent simply by using efficiency measures such as more efficient lighting, cooling, and heating!

Advocates for climate protection definitely have high hopes for these developing countries. In fact, Brazil and China plan to make cuts in their emissions over the next five years that, taken together, will exceed the United States' planned greenhouse gas reductions. The progressive work of these three countries — if it continues at this rate — is at the level needed to help the world level off greenhouse gas emissions in time to avoid irreversible changes.

China

China surpassed the United States as the world's top polluter in the spring of 2007 — an event that many analysts thought would take at least another few years. Despite this dubious achievement, China is taking steps to improve its environmental record. The Centre for Clean Air Policy projects that China's emissions in 2020 will be 7 percent lower than they would have been if they continued to pollute at previous levels had they continued with "business as usual."

China may be improving its environmental citizenship in part because it bore the world's scrutiny as host to the 2008 Olympics. China committed to greening their games in Beijing. They integrated energy-efficient design into all new venues, planted 580 new hectares of forest, and used technologies such as solar power to heat the swimming pools. The government also offset carbon dioxide emissions from the games by investing in projects in China and around the world.

Electric improvements

With its Renewable Energy Law and Energy Conservation Plan in place, China is cutting back on emissions. The Chinese government recently achieved a five-year goal of reducing greenhouse gases from electricity production by 5 percent below their projected levels. This reduction had the same effect as shutting down more than 20 big coal-fired power plants. The government's energy conservation plan also requires the nation's top 1,000 polluting industries to become 20 percent more efficient by 2010.

China is also seeking to develop in a greener direction. The country has been using wind and solar power to provide electricity to remote villages across six regions. Because these areas are remote, it would be difficult to connect them to the power grid. The alternative? Go off the grid and be self-sustaining. This endeavor is part of a five-year project aimed at bringing electricity to 40 million people out of the 132 million that don't have electricity in China. These kinds of projects help build the quickly growing market of renewable energy technologies, making those technologies more accessible and affordable for everyone. (See Chapter 13 for more about renewable energy.)

Changes on wheels

The Chinese government is responding to the country's changing transportation needs. Not long ago, a picture of a Chinese street showed a sea of bicycles. Everyone, it seemed, rode one. Today, that's changing. The *China Daily* newspaper reports that Beijing, China's acclaimed Bicycle Kingdom, now contains up to 2 million cars (although it still has 3 to 4 million bikes) because middle-class residents are changing how they get around.

To deal with this challenge, China's government has implemented vehicle emission standards as part of a National Environmental Friendly Vehicles project that aims to cut emissions by 5 percent below what they otherwise would have been by 2020 — the most rigorous standards in the world. China also taxes cars based on the size of the engine, which means that SUV and truck drivers pay more for their less-efficient vehicles.

The challenges

Even with the steps in the right direction that we talk about in the preceding sections, the growth in greenhouse gases in China is a globally worrying trend. China's carbon emissions increase like they're bringing another coal-fired power plant into operation every ten days. The country continues to use coal and gasoline as its primary fuel sources — two fuels that produce the most greenhouse gas. Coal supplies 98 percent of China's electricity use and 87 percent of its heat energy, according to the International Energy Agency. The use of cars in China is also growing rapidly while people shift from bicycles to automobiles — car ownership has grown 500 percent over the past ten years.

Chinese economic goals are on a collision course with global environmental goals. With China adopting goals to end poverty, and with a growing middle class chomping at the bit to join the lifestyle that rich industrialized countries have created, its emissions are bound to increase unless it shifts its development in a climate-friendly direction. China faces a tough challenge: reducing emissions while becoming an industrialized country.

Brazil

Brazil is caught in a tight spot. On one hand, it has one of the richest ecosystems in the world, the Amazon rainforest, with the Amazon River second in length only to the Nile. The Amazon works as a rain machine, feeding the vast farmland crops that supply food to countries around the world. On the other hand, cutting down the Amazon rainforest to provide land for agriculture is reducing rainfall and creating 75 percent of the country's carbon dioxide emissions (from cutting down or burning trees).

The Brazilian government has programs in place to encourage the use of renewable energy and improve the efficiency of public transit. The government has already committed to deriving 10 percent of the country's electricity from renewable sources by 2022.

Rain reforestation

You can trace the majority of Brazil's greenhouse gas emissions not to cars or industry, but to loss of forests. Tropical rainforests are the most effective of all forests at absorbing carbon (which we talk about in Chapter 3). In the state of São Paulo, for example, only 7 percent of the original forest remains. To deal with this loss of rainforest, Brazil's national government and São Paulo's state government are launching a massive reforestation project, one of the most ambitious reforestation projects in the world.

They don't just plan to plant trees, they also want to rebuild the entire ecosystem: all the flora and fauna found in a rainforest (refer to Chapter 8 for more about ecosystems). Rebuilding an ecosystem is close to impossible, but this project can hopefully restore at least a part of the original rainforest.

Curitiba: A leader in public transportation

One of the keys to reducing greenhouse gases is getting people out of their cars and onto public transit. The Brazilian city of Curitiba is a leader in this area. The concept for that city's system goes back to the early 1970s. The mayor was simply looking for a way to fix two problems — the poor couldn't afford to ride the buses, and the buses needed more passengers. He started a transit voucher system for people who collected recyclables (thus reducing litter) and instructed bus drivers to stop whenever someone waved, regardless of whether they were at a designated stop.

From this humble start, a much bigger dream emerged — to create a city that could integrate work, housing, and recreation to serve Curitiba's quickly growing population. Within three years of its startup, the public transit system served a third of the city's transportation needs. Over the years, the system has added buses and built subway lines, and the city has designed the system in a very efficient way — making sure that buses connect with subway stations and adding express bus lines.

Curitiba's plan has been so successful because it defined from the start how transportation would work in the city and, more importantly, how the city would develop while it continued to grow. It's one of the most substantial long-term transit plans done by a city anywhere in the world.

In a few decades, Brazilians plan to have replaced roughly 1 hectare of rainforest for every 6 hectares lost. This plan doesn't offer a perfect result, but this reforestation is better than doing nothing at all. Community members are involved in reforesting their agricultural lands, making this plan a true grassroots initiative.

Miles ahead (on ethanol)

Brazil has become the world leader in producing sugarcane ethanol as part of its plan to wean the country off oil. Brazil originally focused on ethanol to lower the country's dependence on foreign oil during the 1970 oil crisis — and it worked! Ethanol production has grown into a $65-billion business that makes up 40 percent of the fuel sold in Brazil, according to the Earth Policy Institute. A third of all sugarcane production in Brazil goes straight to producing this fuel.

Rumors often fly that ethanol production in Brazil is leading to more deforestation — but these rumors are exaggerated. Traditionally, Brazil's sugarcane production has happened primarily in the southern region of the county, well away from the Amazonian rainforest. Increasingly, however, ecologists worry that the economic success of sugarcane ethanol may lead to sugarcane replacing rainforest in the northern Amazon.

Jose Goldemberg, a scientist and former Brazilian cabinet minister who has long advocated ethanol, believes that not every country can use ethanol as a solution to reducing fossil fuel dependency — he maintains that ethanol is only one of many pieces needed to solve the climate change puzzle. Goldemberg has said that if even 10 percent of cars in the world were to run on ethanol, it would require ten times the amount of ethanol that Brazil currently produces — an unlikely possibility. (See the sidebar "The man behind the success of ethanol in Brazil," in this chapter, for the story of Goldemberg's contribution to ethanol production.)

But heightened worldwide interest in ethanol makes Goldemberg certain that Brazil will boost production by 50 percent. In Brazil, the majority of cars on the road are *flexible-fuel* — they run on either gasoline or ethanol — and most people choose the more economical ethanol. The government's ethanol program plans to cut emissions from transport by 18 percent below "business-as-usual" trends by 2020, meaning that the emissions from transport will barely increase, even though more cars will be on the road. The bottom line: Sugarcane ethanol in Brazil has reduced carbon output by 9 million metric tons per year.

The man behind the success of ethanol in Brazil

These days, you frequently see ethanol mentioned in the news headlines. But Jose Goldemberg has had ethanol on his mind for over 30 years. He's known as the man behind the global success and acceptance of ethanol as a replacement fuel for gasoline.

Brazil felt a heavy dose of oil shock back in 1975, so, to save its economy, it launched a major campaign on alternative types of oil. Goldemberg was a nuclear physicist at the University of São Paulo at the time and published a paper in the journal *Science* three years after the oil shock. The topic? A message to the world that you can get ethanol from sugarcane — and that Brazil had created this clean and renewable alternative to conventional oil. His work provided the basis for today's ethanol hype.

In 1988, Jose Goldemberg was one of the scientists who attended the landmark first comprehensive international scientific conference on the threat of global warming. The conference, held in Toronto, was entitled "Our Changing Atmosphere: Implications for Global Security." He assisted in drafting the consensus statement and developing a target for reductions 20 percent below 1988 levels as a first step by 2005. He attended the UN's 11th Conference of the Parties in Montreal in 2005, reporting on the agreement between the states of California and São Paulo.

Goldemberg has dedicated himself to energy issues his entire life. His academic work has led him to professorships at universities beyond Brazil — including Stanford, the University of Paris, and the University of Toronto. He published the acclaimed book *Energy for a Sustainable World* (Wiley) over ten years ago, and at the age of 79, he continues to work on developing policy to solve the world's growing energy needs.

A cut below the rest

Deforestation is the biggest roadblock to reducing emissions for Brazil, accounting for 75 percent of the country's annual carbon emissions. While commodity prices for crops go up, farmers want to plant more crops and clear more space for cattle — so, they often remove forests. Seventeen percent of the original Amazonian rainforest has disappeared — an area larger than the size of France.

The rainforest of Brazil is large and difficult to manage — the trees cover an area the size of the whole western United States. No one claims ownership of much of the land, or people are disputing who owns it. So, the Brazilian government can't easily regulate forest activities. Brazil, along with many other developing countries, has argued for including global assistance to stop illegal logging and deforestation within the next phase of the Kyoto Protocol.

The rate of forest loss has been decreasing, dropping 50 percent between 2004 and 2007. The Brazilian government has been working to designate protected forest areas. Thanks to these efforts, 38 percent of *Legal Amazon* (an area that contains all of Brazil's territory in the Amazon basin) is now protected.

India

In terms of its carbon emissions, India is a sleeping giant. According to the World Bank, the country holds 16.9 percent of the world's population but currently accounts for only 5 percent of global carbon dioxide emissions. While the country becomes increasingly urbanized and industrialized, that number could skyrocket — but government agencies and organizations alike are taking the initiative to ensure that India doesn't become a major polluter.

Since 2001, India has spent over 2 percent of its GDP (gross domestic product) on responding to climate change. India has also been the host country for a number of Clean Development Mechanism projects (we talk about these projects in the section "Choosing Sustainable Development," later in this chapter), which have brought about a reduction of over 27 million metric tons of carbon dioxide.

Improving energy efficiency

The Indian government has been working hard at improving the efficiency of non-renewable power providers across the country for the past decade. For example, the government lowered coal subsidies. This loss of funding motivated coal plants to increase their efficiency and even replace some of the coal with natural gas. The government plans to cut national greenhouse gas emissions from transportation and major industry to 12 percent below the projected business-as-usual levels by the year 2020.

Here are some other energy-efficient ventures that the Indian government has undertaken:

- Increasing emphasis on train transportation and shipping, which should boost the country's overall fuel economy
- Converting 84,000 public cars and buses so that they run on compressed natural gas, rather than oil or diesel
- Improving the efficiency of wood stoves in 34 million homes, reducing the number of trees being cut down every year

Big business, big changes

Major businesses in India are getting involved in fighting global warming. And most big banks are coming on board by lending companies money to help cover the initial costs of energy-efficiency projects.

Some projects have shown an immediate payback:

- Apollo Tyres Ltd., a tire production company, invested $22,500 in redoing their heating system, heating the building with the heat from the hot water system. With just a 14-month payback period, the company brings in an annual savings of $19,500.
- Arvind Mills Ltd., also in India, is the largest producer of denim jeans in the world. They connected their two main cooling pumps so that they could turn one off in the winter, a project that saves $280,000 a year — and the project cost them nothing.

A leader in renewable energy

India has become a leader in renewable energy (see Chapter 13). Renewable energy currently supplies 8 percent of the country's total energy needs, and this number is growing. In fact, the country's Electricity Act, enacted in 2003, stipulates that electricity providers in India must derive some part of their energy from renewable sources.

India has invested heavily in wind energy and is now among the top ten wind-energy-producing countries in the world.

India also looks to another developing country for a profitable and carbon-friendly venture. The world's second-largest sugar producer after Brazil, India wants to follow in Brazil's footsteps and start producing ethanol. In 2003, India committed to shifting nine states to a gasoline blend that features 5 percent ethanol, and some states and territories are now moving to 10-percent blends. Though no one knows the exact emission reduction that this measure will create, this move can open up both the Indian and the global market and lower the production price of ethanol.

Growing by the numbers

India's per capita emissions are very low — about one-twentieth of the United States' and a tenth of Europe's. Nevertheless, the country faces substantial pressures to reduce its emissions based on its large population and arguably unsustainable ways. Like China, India has a lot of coal. While India pursues economic prosperity, its growth in greenhouse gas emissions is a serious cause for global concern.

Choosing Sustainable Development

Although we can't overstate the seriousness of poverty and the poor living conditions that many people in developing nations face, these countries do have a tremendous opportunity: Developing nations have the chance to steer clear of the mistakes of the industrialized nations and develop in a way that doesn't harm the planet. Instead of building their nations' economies on carbon-dioxide-emitting fossil fuels, they can choose sustainable development.

What *sustainable development* means is open to debate. The Report of the World Commission on Environment and Development, Our Common Future (also known as the Brundtland Report), which launched the term's popularity, used several definitions. Here are the general concepts:

- Development in both industrialized and developing countries that uses materials in an environmentally responsible way
- Development that doesn't hurt the way that natural ecosystems function; doesn't endanger species; and avoids air, water, and soil pollution
- Development that meets humanity's needs without using so many resources or harming ecosystems to the extent that future generations won't be able to meet their own needs

Climate change and sustainable development are linked. A nation needs a strong and balanced economy (an aspect also in peril because of climate change) to affect sustainable development. Sustainable development promotes

- Renewable sources of energy. These sustainable sources produce low or no emissions, so they don't add to the climate change problem.
- The health and well-being of people, who may be in jeopardy because of the effects of climate change.

Old economies have sacrificed the environment to achieve economic growth. The newly developing countries have a chance to end this historical connection — to develop, but to do so with an eye for the climate change consequences and the incoming effects of climate change.

What developing countries can do

Global warming presents a two-fold challenge to developing countries. While they develop, they need to *mitigate* (or lessen) the production of greenhouse gases. Secondly, they need to adapt to the effects of global warming that they're already feeling.

Mitigation

When it comes to mitigation, developing countries need to *leap-frog* — literally, skip over — the traditional fossil-fuel-based model that the older industrial economies followed and move straight into renewable energies to avoid boosting greenhouse gases. They have to choose new ways of generating energy, such as using solar, wind, low-flow tidal, and geo-thermal power, as well as bio-fuel. (We talk about these alternate energy sources in Chapter 13.) Developing countries can't always easily get these cleaner, renewable technologies, but the technologies have numerous benefits. Improvements in the energy sector also improve community health and general productivity, and boost the economy.

Clean energy can be costly, particularly for poor nations, so the Kyoto Protocol (a global climate change agreement) includes a program called the Clean Development Mechanism. (Refer to Chapter 11 for more about the Kyoto Protocol.) The Clean Development Mechanism (CDM), which the Brazilian government initially proposed, emerged from the last round of negotiations to create the Kyoto Protocol. In this program, developing and industrialized countries agree to work together, and the industrialized countries pay to build clean energy projects in the developing countries. (We talk about this program in the section "How industrialized countries can help," later in this chapter.)

One example of a successful CDM project is in Honduras, where a power-generating company collects the waste from a palm-oil mill. The company uses this plant-based waste as biogas to generate electricity. Before the company collected the waste, it sat in ponds while it decomposed and gave off methane. This project uses the waste, thereby reducing the methane emissions. Capturing the methane can reduce over 156,000 metric tons of greenhouse gas emissions each year, and the biogas energy can go straight into the grid to offset a total of over 37,000 metric tons of emissions in those same seven years that the palm oil waste decomposed in the ponds.

Adaptation

Although developing nations have the chance to avoid the mistakes made by industrialized countries, they can't avoid the consequences of the mistakes already made. Developing nations have been feeling the effects of global warming for over a decade. The impacts of climate change are likely to worsen, so developing nations must adapt to prevent avoidable loss of life and property. (We discuss the disproportionate effects of climate change on developing countries in Chapter 9.)

Adaptation takes many forms:

- Shifting to more drought-resistant crops.
- Rethinking transportation infrastructure to locate bridges away from areas vulnerable to flash floods.
- Protecting and rehabilitating mangrove forests, which can help protect coastlines from increased storm surges.
- Moving vulnerable populations away from low-lying islands or coastal zones that may become flooded out of existence.

Some countries need to take all these measures — and more.

One major adaptation project that developing countries can undertake is planting trees in areas suffering from deforestation. Planting trees not only cools down the planet (a positive characteristic of trees and plants that we talk about in Chapter 2), it also enriches the soil with nutrients and helps reduce water runoff from flash floods. The soil on a treeless hillside washes away in a mudslide, but a tree-covered hillside's soil stays put. The fewer natural disasters in a developing country, the less damage the local community and economy suffer.

Reforestation projects are taking place all over Brazil. A company doing the damage in the first place actually heads one of them! Performance Minerals & Pigments cuts down large strips of forest every year while they mine for a rare clay that's used to make the glossy finish on paper. They've committed to reforesting the entire 10,000-hectare site that they've stripped.

Sadly, adapting to global warming also means preparing for the worst. Developing countries need safe water and food supplies to ensure that people have enough to survive if one of the extreme weather events — such as hurricanes and floods — wallops the developing world. To guard against those storms, governments need to build higher dikes and make stronger bridges. They need to make riverbanks and seasides capable of coping with deluges. They also need to make plans for a world with too little water; already, the World Bank is working with many other development agencies around the world, such as its work with several Caribbean countries to develop drought-resistant crops, while helping Bangladesh plan ahead for serious loss of territory due to sea level rise.

Adaptation also means adjusting to whatever changes climate brings. Some communities may need to relocate if they can no longer sustain themselves. If a flood washes out a road or a hurricane levels a town, you shouldn't rebuild in the same way or in the same locations.

Endangered nations

In 1987, at the General Assembly of the United Nations, President Maumoon Abdul Gayoon of the Maldives warned that the world would soon have to use a new term. Humanity would talk about not just endangered species, but endangered nations. He urged that the world's wealthiest nations act on global warming.

Ten years later, President Gayoon spoke at the United Nations to mark five years since the Rio Earth Summit, saying that the world had ignored the pleas of low-lying island states. As a result, the Maldives was forced to relocate villagers, as well as the population of an entire island, to higher ground.

How industrialized countries can help

Industrialized countries have two roles to play in helping developing countries adapt to climate change — as leaders and as partners.

Industrialized countries need to take the lead in clean technologies to show developing countries that the industrialized world is dedicated to cutting greenhouse gas emissions and that it recognizes its part in creating most of the emissions to date. (We talk about the actions that wealthy nations can take — both cutting emissions and adapting to climate change — in Chapter 11.)

Beyond developing and sharing low-carbon technologies, the United Nations Framework Convention on Climate Change (UNFCCC) says that industrialized countries need to partner with developing countries to help them cope with climate change. Industrialized countries have both the resources and the responsibility to work on these projects.

Aiding developing nations through the Clean Development Mechanism

The Clean Development Mechanism, or CDM (part of the Kyoto Protocol, which we talk about in Chapter 11), is one program that has developing and industrialized nations already working together to reduce greenhouse gas emissions. Through this program, industrialized countries fund greenhouse gas–reducing initiatives in the developing world. Smaller players usually implement the initiatives — a local renewable energy company, for example, or a municipality that has applied for support for the project.

Some industrialized countries become involved in CDM because of political will — the diligence to do what is right despite any barriers. Others, however, might invest in CDM projects to offset the weakness of their own climate change plan; investing in CDM projects increases a country's carbon credits, which help the atmosphere in any event by reducing carbon emissions, even if from another country.

The United Kingdom currently supports over a third of the CDM projects in the world. Japan and the Netherlands each support about a tenth of the projects.

One innovative feature of the CDM is that every CDM transaction includes a tiny tax. Every time a wealthy country invests in a poor country for a carbon-reduction project, the trade includes a surcharge of 2 percent that goes to a fund for adaptation. This first global tax to fund adaptation is a step in the right direction, but it has nowhere near the resources that a proper global adaptation effort needs.

A similar financial mechanism is required to generate the tens of billions of dollars needed every year to fund adaptation in developing countries. This could be a tax, similar to that on CDM transactions, but for all exchanges of emission credits. Another mechanism of similar scale is also needed to allow developing countries to access clean energy technologies, allowing them to lift their citizens from poverty without following industrialized countries' dirty path of development. Such mechanisms are being discussed as part of the UN negotiations for the agreement to follow Kyoto.

Planning on it

The world can no longer avoid some changes caused by global warming. So, all countries need to come up with a plan for how to deal with and adapt to climate change — a plan that countries actually use and update on a regular basis. While the climate changes, so must the plans and strategies of countries around the world.

Making adaptation plans for climate change, which is so diverse in its effects, means dedicated cooperation between government departments and across sectors of government, industry, business, and the public. Regular communication between different regions helps coordinate the planning and enables the government to make adjustments to the plan based on successes and failures.

For developing countries, plans for adaptation must include stable funding. This funding can come directly from industrialized countries or from organizations funded by those countries — such as the Global Environment Fund (GEF). The GEF deals with a range of environmental issues, including climate change. It gives grants to projects in developing countries around the world. So far, they've dispersed $6.8 billion since 1991, helping over 1,900 projects.

Unfortunately, the world has made fewer meaningful efforts on adaptation, particularly for the developing world, than its efforts to reduce greenhouse gases. "Too little, too late" is not the epitaph humanity wants! The world has to deal with climate change right now, and every nation needs to plan for today, tomorrow, and well into the future to literally live with the changes.

The UN has registered over 1,000 cooperative projects and expects that, by 2012, all the projects will have reduced emissions by almost 3 billion metric tons. The bigger projects have produced reductions of over 10 million metric tons of carbon dioxide, with the smaller projects accounting for reductions of 500 to 1,000 tons each year. Just a few of the big projects include the following partnerships:

- A private power plant in India and a private carbon-management firm in Switzerland are building a biomass-fired co-generation plant to power an industrial facility in Koorghalli Village in India (see Chapter 14 for information about co-generation). The total emission reduction will be over 1 million metric tons in the project's seven-year period — reducing over 36,000 metric tons every year (equivalent to taking 12 million cars off the road every year).
- A company from the Philippines has paired with a company in the United Kingdom to capture methane gas from landfills. The Montalban Landfill Methane Recovery and Power Generation Project, which uses the methane gas to create energy, will cut a total of 5.8 million metric tons of greenhouse gas emissions.
- The Netherlands and Sweden are partnering with wind-power-generation companies to increase the wind-power capacity of the Ningxia region of northern China, feeding that wind power into the electricity grid. This project is saving almost 100,000 metric tons of carbon dioxide a year, not to mention providing employment opportunities and helping reduce poverty in the area.

Both the private sector and governments are developing more projects because interest is growing. While partnerships grow and companies build on experience, these projects in developing countries can improve and therefore have greater impacts.

Acknowledging an imbalance in Africa

Clean Development Mechanism projects are concentrated in Central America, Brazil, China, and India. Very few CDM projects are underway in African countries for the following reasons:

- Africa is a relatively clean continent relative to the others, so it has fewer opportunities and reasons for industrialized countries to invest in projects there.
- Countries must have the people and the organizational capacity to support a project. Projects require government cooperation, and governments require the capacity to assist the projects. Additionally, the country's infrastructure must be able to support the project. Unfortunately, many African countries lack these qualifications.

- Investors look for low-risk projects — mainly in countries that have little political and economic instability. Many countries in Africa are less stable than countries such as India or China.

Nevertheless, the potential for low-carbon development strategies is very high for African nations. The Prime Minister of Ethiopia has said that he would love to see his nation become carbon-neutral. He lacks only the investment partners to make it happen.

The United Nations is trying to address the imbalance of more development strategies in Latin America, China, and India versus Africa, but a gap still exists between these international development agencies and the communities in developing countries who must ultimately adjust the balance themselves. Although the emission reduction projects that pose the biggest opportunity for CDM credits focus on lowering the emissions of a power plant or changing the source of electricity delivered to a community, African countries may find *afforestation* — planting trees where none grew before — a more accessible option.

While the Kyoto Protocol develops and improves, so can the CDM program. The UN negotiations created The Nairobi Framework in 2006, which is committed to helping bring more CDM projects to Africa.

Part V

Solving the Problem

In This Part . . .

Can humanity actually avoid getting to the point of huge, devastating, and irreversible changes in the world's climate? Of course! (Insert sigh of relief here.) A few conditions are already set in stone with Mother Nature, but people still have time to veer away from the most disastrous impacts.

This part is the most exciting because it's packed with solutions. We look at renewable energy sources, such as wind and solar power, and consider what industry can do to become climate friendly. Another important solution involves spreading the word about global warming, and keeping businesses and governments informed about the problem, so we fill you in about what non-governmental organizations are doing and how you can get involved. We also offer information about how to stay on top of media reporting about global warming. Finally, we fill you in on what you can do to reduce your own carbon footprint, whether you're traveling, at home, or at work.

Chapter 13

A Whole New World of Energy

In This Chapter

- Relying less on oil
- Burying the emissions from oil
- Comparing renewable energy sources
- Looking at the upside and downside of nuclear power

Citizens around the world rely on oil and other fossil fuels as their major source of energy, but times are changing. Oil — at least cheap, accessible, and conventional oil — isn't as abundant or secure as it used to be. Global warming has awoken people to the dangerous greenhouse gas emissions that burning fossil fuels produces. Humanity needs to change the way it uses energy and where it obtains that energy.

Renewable energy is a secure energy source, helps reduce the use of fossil fuels, and gives people a real option for cutting greenhouse gas emissions. In the next 25 years, the International Energy Agency (IEA) projects that the global use of wind, solar, geothermal, and hydroelectric power will continue to grow. But it acknowledges that although renewable energy *can* entirely replace fossil fuels, it likely *won't* within the next 50 years. Others, such as Dr. Hermann Scheer, architect of Germany's Renewable Sources Act and the leading proponent for the creation of a United Nations–sanctioned International Renewable Energy Agency (IRENA), argue that not only is a significant shift towards 100 percent renewable energy possible, but it is absolutely necessary.

In this chapter, we look at how the world's use of oil and other fossil fuels needs to change, and we explore the opportunities for renewable energy.

Changing the Way Civilization Uses Oil

Changing habits and technology is no easy feat, but it's something everyone will be doing in the coming years. The world, especially industrialized countries, has begun to think about oil as a precious resource that must be managed carefully and used frugally.

Becoming conservative with oil use is important because demand is about to surge: The International Energy Agency (IEA) expects world energy demand to rise more than 50 percent in the next 20 years due to rapid industrialization in developing countries.

Prioritizing and conserving fossil fuel use

Because modern civilization's love affair with oil has a time limit, people need to make that relationship last — and to make a few compromises along the way. Humanity needs to start using oil only where it's really needed and replacing it with alternatives where they exist.

The economic activities for which *only* petroleum products can meet current needs lie largely outside the energy field. Petrochemical production includes everything from plastics and nylons, synthetic fabrics, and a host of consumer products. Burning petroleum, with its many other uses and its limited supply, is hardly the most sensible approach.

The modern world now needs to play the field and add other energy sources into the mix where fossil fuels can be reduced or replaced. You may have already seen gas that's been blended with ethanol or diesel that's been combined with biodiesel at the pumps. Because these fuels use up to 25 percent less gasoline and 10 percent less diesel, respectively, you can help conserve fossil fuels by using them. (We discuss these fuel alternatives in the section "Investigating Renewable Energy Options," later in this chapter.)

In some instances, people can side-step the use of fossil fuels completely. Alternative forms of car fuel and energy to power buildings and industry have been around for a long time — ranging from electric cars to renewable energy sources. Even some alternatives to consumer goods are readily available. Plastics can be made from corn oil. Fleece and other synthetic materials can be made from recycled plastic bottles. Other alternatives are in the planning stages: One example involves a possible solar-powered airplane — the sky's the limit!

Combining heat and power

When oil, gas, or coal is burned to generate electricity, only about one-third of the energy actually generates any power. The other two-thirds is wasted. A lot of it goes up the chimney. (This type of energy waste is more common than you might think. With regular light bulbs, only 10 percent of the energy creates light — the other 90 percent is lost as heat!)

This lost energy doesn't have to go up in smoke; it can be used for *co-generation*, producing both heat and electricity at the same time. Co-generation takes the same amount of fuel, but it captures the heat that would be lost otherwise and uses it to heat a building. COGEN Europe reports that co-generation systems capture about 90 percent of the energy this way, with only a scant 10 percent being wasted. New York City's Consolidated Edison utility uses the steam from seven steam generating plants to provide heat to 1,800 commercial customers, who use it to warm 100,000 buildings.

Co-generation technology can also be used with renewable power plants that use biomass (like forest waste) or biogas (produced by using bacteria to eat up cow manure and food wastes) and can be supplemented by solar hot water heaters that pre-heat the water used to produce steam in the generators, meaning fewer fossil fuels (or bio-fuels) need to be used.

Figure 13-1 shows a co-generation system that uses turbines to reuse energy from pressurized blast furnace gas.

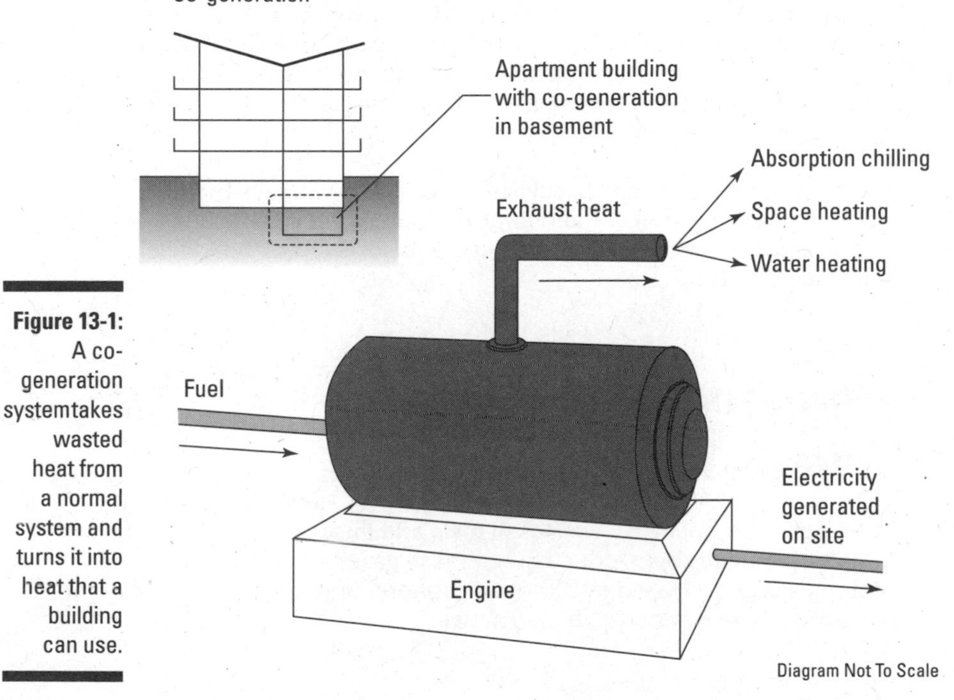

Figure 13-1: A co-generation systemtakes wasted heat from a normal system and turns it into heat that a building can use.

Using oil efficiently

When no current option exists besides using fossil fuels, people can at least ensure that they use it efficiently, squeezing as much energy out of it as possible. Companies are working to develop high-efficiency technologies. For example, the compact fluorescent light bulb, which lasts ten times longer than a regular light bulb and uses a quarter of the energy, has already found its way into many homes. (Although these bulbs use electricity, fossil fuels often generate that electricity; refer to Chapter 4 for more on how people use fossil fuels.)

Just about anything — whether it's a whole office building, an entire fleet of cars, or a total manufacturing system — can be made more efficient. People often use four to ten times more energy than they need, according to the European Renewable Energy Council. It doesn't make sense, for example, to use oil or grow fields upon fields of corn for ethanol when someone could simply redesign cars so that they use ten times less fuel to begin with.

Civilization was improving its efficient use of energy at a rapid rate until 1990. Since then, that progress has fallen off; the International Energy Agency (IEA) reports that energy efficiency has been improving at only half its previous rate. Humanity needs to get back on track and double its rate of energy-efficiency improvement if it wants to make a significant difference in the fight against global warming.

Technology hasn't yet hit its efficiency peak. The IEA says that a lot of room still exists for development and improvement of high-efficiency measures in industrialized countries, especially in the major sectors: buildings, industry, and transportation.

Changing How to Handle Fossil Fuel's Emissions

Even if civilization uses fewer fossil fuels and improves how efficiently it uses it, fossil fuels will still produce greenhouse gases when burned. Currently, those gases are released into the atmosphere, but oil companies and researchers are exploring an alternative.

What's a fossil fuel company to do?

Oil companies face huge opportunities. Consider British Petroleum (BP), which now markets itself as Beyond Petroleum. Responding to climate change, BP has taken social and environmental leadership to minimize its emissions and overall environmental impact. These changes were sparked under the leadership of former Chief Executive Lord John Brown. What was once strictly a petroleum business has morphed into an energy business that raises public awareness about climate change and big technology solutions.

BP still deals with petroleum, but it also takes a lead in developing low-carbon fuels and technology. The company

- Launched an alternative energy business in 2005
- Launched a bio-fuels business in 2006
- Invested $520 million in projects researching low-carbon technologies, clean energy, energy policy, and urban energy solutions
- Is involved in developing carbon capture and storage technology
- Is dedicated to raising public awareness through advertising and Web site features

Fossil-fuel companies have the potential to become energy companies, expand their customer base, and ensure that they remain successful through policy and cultural changes designed to help reduce greenhouse gas emissions.

Capturing and storing carbon dioxide

Oil companies and researchers are conducting major tests dealing with the capture and underground storage of greenhouse gases. Storing the carbon works differently in different places. Two ways of capturing and storing carbon are currently being used:

- **During oil production:** Oil-product producers pump the carbon dioxide into the ground at the same time that they pump the oil out of the ground. The pressure from the gas being pumped in actually helps to get the oil out more efficiently — creating a controversy over whether this use of the technology actually benefits the climate. (See the sidebar "Carbon capture controversy," in this chapter, for more information.)
- **Dealing with industrial emissions:** Major carbon dioxide emitters, such as coal power plants, capture the gas by containing the source of the emissions and directing the gas underground. This process stores the carbon dioxide in places from which people once extracted oil; in big, empty spaces underground; or into unmineable coal beds and saltwater aquifers. This need for storage-space limits the widespread use of this technology.

Figure 13-2 shows how both of these methods work.

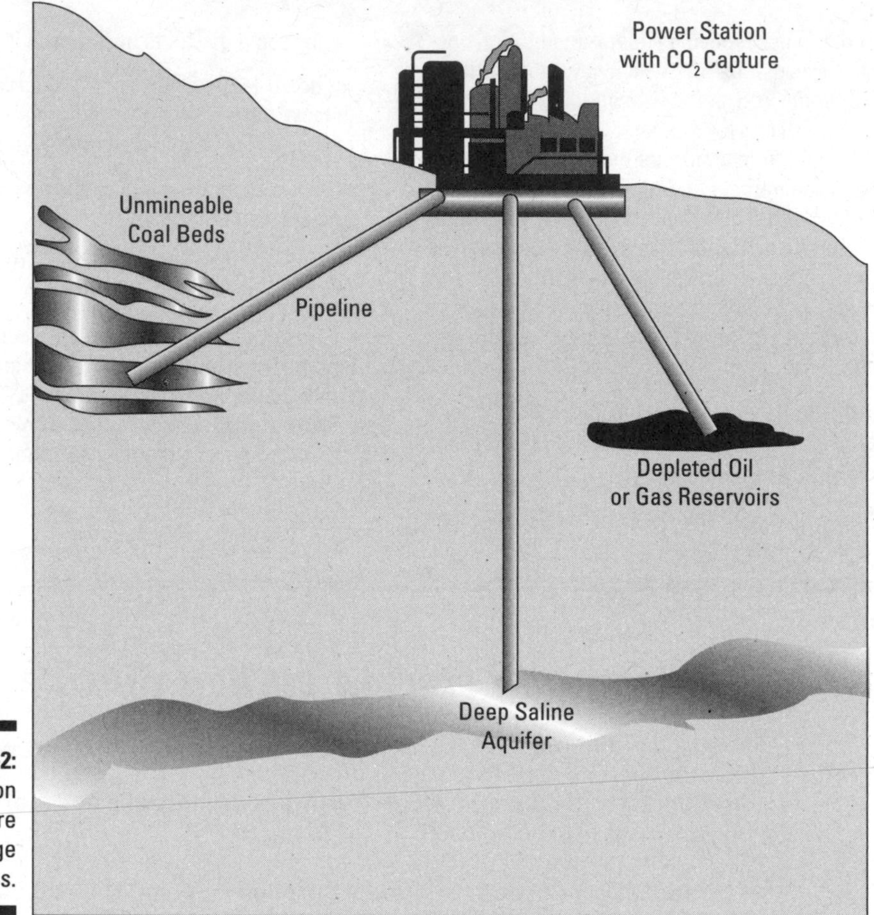

Figure 13-2: How carbon capture and storage happens.

An area's geology is a huge factor in determining whether or not carbon capture and storage will work there. In Norway, for example, the technology is being applied by pumping carbon dioxide down under the deep sea. But what works in the North Sea geologically may not work elsewhere.

People can't capture all carbon dioxide emissions. Catching the emissions coming out of a stationary smoke stack is much easier than capturing the carbon blowing out of the tailpipe of a moving car. Over half of the carbon dioxide that can be captured is emitted by coal power plants; the remainder is from major industrial emissions.

Carbon capture controversy

In Canada, the Weyburn Saskatchewan Project is capturing the emissions from the Great Plains plant in Beula, North Dakota. The carbon dioxide emissions are transported through a 202-mile (325-km) pipeline and stored in old oil fields at Weyburn.

This project plans to store up to 20 million metric tons of carbon dioxide — while producing another 130 million barrels of oil. Some argue that they wouldn't be able to access that much oil without the carbon dioxide; so, the entire process may ultimately produce more carbon dioxide than if none had been captured at all.

Considering carbon capture cons

Carbon dioxide capture and storage is a temporary solution because the available storage space is limited. (This storage space could also be employed for other, greener uses including the storage of natural gas and the compression of air as a means to store excess energy that can be released when necessary to turn turbines.) Whether people can pipe carbon dioxide back underground, down the now-empty oil wells, depends on the kind of geology near the pollution source. According to the International Energy Agency (IEA), the planet has enough room underground to store tens to hundreds of years' worth of carbon dioxide emissions (no one knows just how much room exists, hence the wide time range).

Some environmental groups have major concerns about carbon capture and storage, arguing that it

- Allows civilization to keep using fossil fuels, rather than replacing them with carbon-friendly alternatives.
- Enables oil producers to extract more oil, ultimately creating more carbon dioxide emissions (see the sidebar "Carbon capture controversy," in this chapter, for more).
- Isn't foolproof. The stored carbon dioxide has the potential to leak out of storage and escape into the atmosphere, thus warming the atmosphere. Additionally, the escaped carbon dioxide, which is odorless, colorless, and poisonous, could collect in hollows, posing a threat to animal and human life.

Despite challenges and concerns, the IEA predicts that carbon capture and storage could be ready for major use in industrialized countries by 2015 and play a huge role in reducing worldwide greenhouse gas emissions by 2050.

This technology is still expensive, but its costs could come down by half or more if initiatives such as the carbon market are implemented. (See Chapter 10 for more on the carbon market.)

The cost of carbon capture and storage isn't cut and dried — it depends on factors such as how far the carbon dioxide has to be transported for storage and what kind of technology is used to store it. The estimates for capturing the gas range from $5 to $115 per metric ton of carbon dioxide — depending on what technology is used. Transporting the gas can cost $1 to $8 per metric ton. Storing it can cost 50¢ to $8 for under the ground, and $5 to $30 for under the ocean floor (the ocean-floor storage costs more because carbon-storing companies have to transport the gas about another 60 to 300 miles — roughly 100 to 500 km — offshore).

Investigating Renewable Energy Options

The threat of climate change has encouraged individuals, governments, and businesses to consider sustainable energy sources that have low greenhouse gas emission rates, rather than exhaustible fossil fuel resources. Sustainable, or *renewable,* energy sources — such as plants, the wind, and the sun — regenerate themselves or can be replenished. Once up and working, most of these energy sources produce zero or little greenhouse gas — the only emissions from wind and solar power, for example, could come from the manufacturing of the turbines and panels, not from the actual production of electricity.

None of these sources can supply all the world's energy needs on its own, but collectively, renewables can replace fossil fuels. The primary barrier has been the cost of renewables versus fossil fuels. The solution lies in choosing which energy source is right for a particular region. For a secure and sustainable future, humanity needs a range of energy sources, making civilization's energy systems less vulnerable to changes — like having a bunch of backup systems.

Renewable energy is already used around the world. Renewable sources make up 4.9 percent of the energy produced in Australia, 3.8 percent of Canada's energy, 0.6 of the United Kingdom's, and 3.9 of the United States'. The global average is a bit lower, at 3.5 percent. Those figures are much higher if you include large-scale hydroelectric systems, such as northern Quebec's James Bay facilities. (The percentages that include hydroelectric power are 5.5, 16.1, 2.0, and 4.7, respectively.)

Calling hydroelectric systems "renewable" is controversial because water may not always flow where it does now. Calling these projects "sustainable" also draws some controversy because they can permanently inundate thousands of square miles of territory for enormous reservoirs.

Blowin' in the wind

With the power to uproot trees, tear apart homes, and transport Dorothy to Oz, the wind is a natural force to be reckoned with. Humans can harness that power to generate energy through overgrown pinwheels, known as wind turbines. A *wind turbine* typically features a propeller that has three blades on a hub that sits atop a tall pole. When the wind blows, the blades of the propeller spin, capturing the wind's power and transforming it into electricity or mechanical energy. Together, multiple turbines are called a *wind farm*. Figure 13-3 shows a wind farm.

Figure 13-3: A wind farm in California.

Glen Allison/PhotoDisc/Getty Images, Inc.

Worldwide, the wind-power industry has been growing at 23 percent a year since the early 1970s. And, while the industry grows, wind power technology has become more reliable and cheaper.

Three countries boast two-thirds of the world's active wind power: Denmark, Germany, and Spain. Denmark has long been the leader in wind power development, with 18 percent of its energy coming from wind in 2005, according to the IEA. Wind energy has a significant upside: It produces zero emissions. The IEA expects that building up the wind industry can help make huge reductions in greenhouse gas emissions.

Unfortunately, a few barriers prevent the widespread implementation of wind technology:

- **Costly to connect:** Good sites for wind power are often far from transmission lines, and installing new lines is difficult and costly. Some wind farms are now generating hydrogen, which can be delivered along the same kind of systems used for natural gas.
- **No storage:** Currently, wind power can't be stored effectively, so it's available only when the wind blows. Countries like the Netherlands, Germany, and Spain are beginning to utilize different forms of storage such as compressed air, pumped storage (taking water to the top of a hill to use later), and even hydrogen production.
- **Public opposition:** Public acceptance can still be low, depending on where builders place the wind farms. People don't often want to have wind farms in their communities because they think those farms are unsightly. Wind farms also produce low-level frequencies that some people find unpleasant. Finally, some people worry that wind farms endanger birds. (For more, see the sidebar "A big flap over wind power," in this chapter.)

A big flap over wind power

One of the biggest complaints against wind turbines is that they kill birds. Initially, this threat was a serious problem because builders didn't take bird and bat migration routes into consideration when erecting the turbines. A report put out by the U.S. National Academy of Sciences says that for every 30 wind turbines, one bird is killed each year, although these numbers are contested. Based on these numbers, 40,000 birds die because of wind turbines each year in the U.S. alone.

While wind energy moves forward, wind turbine developers and companies are taking concerns for bird safety into account. They usually build turbines away from migration routes and often add colored markings to the blades to ensure that our feathered friends can see them. Ultimately, wind turbines are far from a bird's worst enemy, making up just 0.003 percent of human-caused bird deaths in the U.S.

Here comes the sun

Solar energy is very efficient and, like wind power, it produces no greenhouse gases. You can capture solar energy, even on cloudy days. And it's the most abundant type of energy available — far beyond civilization's needs. In fact, the World Energy Council reports that the amount of solar energy that reaches the earth is over 7,500 times that of civilization's current energy demand! Even if society only captured 0.1 percent of the sun's energy, and used it with just 10 percent efficiency, it could still meet its current energy demand four times over.

The downside, which many renewable energy sources share, is the high upfront cost of setting up new technologies to capture that energy. Presently, solar electricity generation is the most expensive form of renewable generation (while solar thermal energy for producing hot water is one of the cheapest). Oil and coal offer far cheaper alternatives, but at a great environmental cost. Meanwhile, fossil fuel prices are going up, while solar costs are dropping.

People can harness solar energy in three ways, each of which we investigate in the following sections.

Photovoltaic energy

When people think of solar power, they probably think of *photovoltaic* power sources, which use solar cells to convert sunlight into electricity (*photo* means light, and *voltaic* means voltage). You can use these solar cells on their own — to power things such as calculators, satellites, or flashing electric construction signs — or as a part of a power grid, in which they contribute to an area's energy source.

One of the most common ways to use this type of solar energy is on roofs — using solar panels, solar shingles, solar tiles, or even solar glazing on skylight windows. You can have them designed for any size building with varying energy needs. When Bill Clinton was president of the U.S., he set a goal of a "Million Solar Roofs." This effort helped the photovoltaic industry get off the ground in that country.

The one drawback to photovoltaic power sources is that the typical solar cell is relatively inefficient. For it to work at its best, it generally needs direct sunlight — though it does still work on cloudy days. Fortunately, solar energy developers are continually working to improve this efficiency.

While the use of photovoltaic power spreads, the price comes down. Japan, Germany, and the U.S. lead the world in widespread solar cell use. Some solar companies are now boasting that their sun-made electricity can cost-compete with coal.

What's a watt?

Power is measured in *watts* (W), named after steam-engine inventor James Watt. It refers to an amount of energy consumed at a point in time. Lifting an apple off the table takes one watt.

You probably know the term "watts" from buying light bulbs of 60 W or 100 W. When a 100-watt bulb is on for ten hours, it uses 1,000 "watt-hours" (the 100 watts of power it draws multiplied by the ten hours it's on). This amount of energy is more commonly called one kilowatt-hour or kWh. The electricity for your home is normally billed by the kilowatt-hour. Depending on how energy-efficient your fridge is, running it for a year uses 400 to over 900 kWh of electricity.

Solar photovoltaic energy tends to produce the most power when the demand is highest (on hot days in the summer when people pump up their air conditioning). Because electricity demand is higher at these peak times, the price of electricity is very high, making solar power very competitive and useful — especially if it's being produced close to where it's being used (like on your roof) so that as little precious power is wasted as possible (transporting power over the lines uses up power).

Passive solar energy

You can use the sun as an energy source, even without special technology, simply by using its direct heat and light. This type of solar energy is called *passive* solar energy because you don't need to physically transform the energy to use it. For example, the heat from the sun coming through your kitchen window warms up the room. (Cats are famous for taking advantage of passive solar energy.)

This technique works because it follows the sun. The sun rises in the east and sets in the west. It is also higher in the sky in the summer, and lower in the winter. Using passive solar energy means shading yourself from the summer sun, and bringing in the warmth of the winter sun.

For example, to take advantage of passive solar energy in your home, you can install most of your windows so they face south (north if you're in the southern hemisphere) and fully insulate the opposite side of the house. You can also put your most-used rooms, such as the kitchen and living room, on the south side of the house and put your bedrooms on the opposite side. With this kind of house design, the sun's warmth heats and lights the kitchen and living room during the day. This warmth eventually heats the whole house. At night, your well-insulated bedrooms keep in the heat that built up over the day. In the summer, the sun's rays from above will hit the roof, and not the windows, keeping your home cool in the summer.

Solar thermal energy

Solar thermal energy is sort of a combination of photovoltaic and passive solar energy: It uses panels that have water or *glycol* (an antifreeze-type liquid) running through them to capture the heat from the sun. You can use this heat for different purposes, depending on the solar thermal system.

The most common use for solar thermal energy is to heat things such as swimming pools (by using low-temperature collectors) and residential hot water tanks (by using medium-temperature collectors). Medium-temperature solar thermal energy collectors can heat more than water, however; they can even heat large commercial or industrial buildings. Take a look at Figure 13-4 to see how solar thermal energy works.

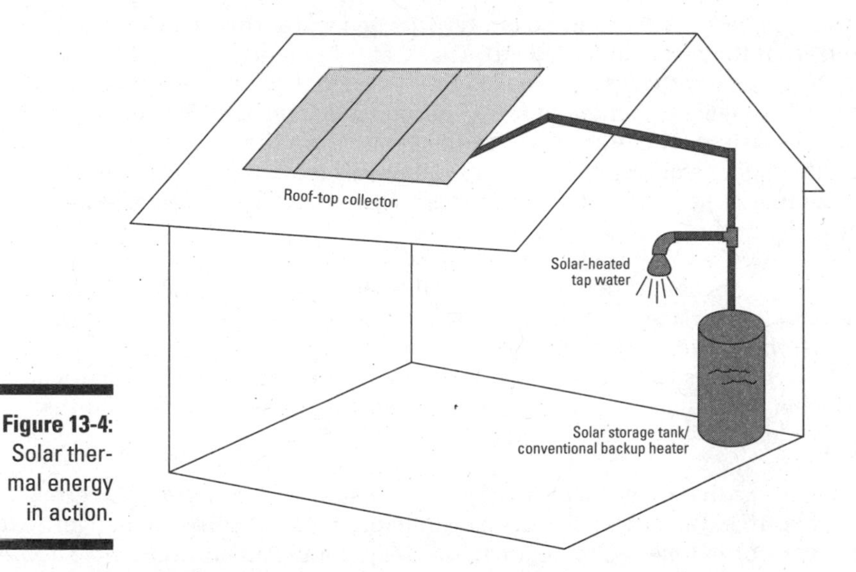

Figure 13-4: Solar thermal energy in action.

You need other means of energy to heat water — solar heat can't do it all. Depending on how sunny it is and how much hot water you use, solar hot water heaters can supply, on average, 60 percent of the heat you need for your hot water, according to the National Renewable Energy Laboratory. Surprisingly, 50 percent of homes in the U.S. and 67 percent of commercial properties are in locations with sufficient sunlight to use solar thermal energy.

China, Taiwan, Europe, and Japan use the majority of solar thermal energy, and they use it for hot water and space heating. In Canada and the U.S., the technology is mostly used to heat swimming pools, though more homeowners are getting on board (especially when they find out they can pay the system off in five or six years if they have a family of four or more).

You can also use solar thermal energy to produce electricity. High-temperature collectors, which take the form of multiple mirrors, concentrate sunlight and focus that energy on a small area. The resulting heat can produce steam from water, which you can use to generate power.

Heat from the ground up

You know the Earth is heating up — that's why you're reading this book. But beneath the Earth, it's already hot. *Geothermal* power derives from the heat beneath the Earth's crust. (*Geo* literally means Earth, and *thermal* means heat.)

People can harness geothermal energy through water (or an antifreeze-type liquid) that they pipe underground. That water boils because of the heat from *magma* (what volcano lava is before it reaches the surface) deep inside the Earth. Magma is incredibly hot — around 1,800 degrees Fahrenheit (1,000 degrees Celsius). The pressure from the steam from the boiled water drives itself through installed pipes back up to the surface and propels a turbine, which creates electricity. The pipes then return the water underground to repeat the cycle.

Some hot spots are more readily accessible than others. The closer a hot spot is to the surface, the easier it is to access. Also, the easier it is to get through the ground, the better chance you have of accessing a hot spot. The United States has the most geothermal resources, with Latin America, Indonesia, the Philippines, and East Africa also well endowed. Geothermal energy plants exist on every continent in the world.

Even in areas that do not have access to "hot spots," geothermal is a growing energy source. Water can be warmed sufficiently using a heat exchanger (like a refrigerator in reverse) to heat a house by pumping an antifreeze-like liquid through a closed loop underground (a loop of pipes buried in your back yard or under a parking lot). A simple geothermal heat pump can be located nearly anywhere on earth. This form of geothermal energy does not make electricity, but it can replace a fossil fuel furnace.

About 19 percent of Iceland's electricity comes from geothermal energy. The country has 600 underground hot springs that it can tap into, enough to fuel all of Iceland's electricity. The water itself heats about 90 percent of homes in Iceland and provides all the hot water. Geothermal utilities use the same source to create the many bathing pool hot springs in the country. Iceland has been shifting its electricity technology from oil to geothermal ever since the oil crisis in the 1970s — they've invested about $8 billion over the past three decades, and they've become almost entirely self-sufficient. The Philippines are close behind Iceland, generating over 17 percent of their electricity from geothermal energy.

The IEA estimates that the world offers a potential of 85 gigawatts of geothermal energy (a *gigawatt* is 1 million watts), which is 0.6 of a percent of humanity's current energy demand. Geothermal energy sources currently supply the world with more energy than solar and wind energy sources combined.

Geothermal power does have its downsides. Installing a system takes a long time, and the drilling is costly — similar to drilling for oil, but without as big a financial payback. Additionally, concern exists that drilling for geothermal energy can cause earthquakes because every geothermal hot spot is in a geologically active area. Supporters of geothermal energy contend that these quakes would likely be so small that you wouldn't be able to feel them. Others maintain that not enough evidence exists to say that geothermal drilling causes earthquakes at all.

Another way to take advantage of our planet's underground warmth is through *earth energy,* which doesn't take its heat from magma, but from natural underground hot springs. Pumps can tap into these waters and use the warmth to heat water or the interior of a building. We discuss these heat pumps in Chapter 18.

Hydropower

Anyone who's been to Niagara Falls can attest to the beauty and majesty of rushing water. It's also a great way to generate energy. *Hydropower* uses the flow of water to turn turbines, which convert the energy into electricity. This method is very similar to how a wind turbine generates electricity.

People can generate hydropower in two ways:

- **Impoundment systems:** Water is stored in dams and reservoirs, which hydropower utilities can then release to help meet power demand at specific times.

  Unfortunately, building large dams can flood natural ecosystems and even nearby communities. The IEA also reports such problems as increased fish deaths and land erosion. (For an example of a problematic impoundment system, see the sidebar "Dam it," in this chapter.)

- **Run-of-river hydropower plants:** Water's natural flow is used to produce power continuously. The huge power plants at Niagara Falls, for example, are run-of-river. If you have a stream or river on your property, you might be able to use it to provide some or all of your own house's electricity needs. The run-of-river hydropower systems cause very little environmental damage.

Dam it

Looking over the Yangtze River, the third longest river in the world, Chinese developers (envisioning a huge source of hydropower) said, "Dam it." Under development when we wrote this, the Three Gorges Dam on the Yangtze River is a 15-year, $25 billion project that will be the world's biggest hydropower dam by the time it's complete in 2009.

The dam's impact on locals and on the environment has been widely criticized. The dam has already displaced 1 million people, the majority of whom are poor, and critics say the government hasn't properly resettled the displaced. When it's complete, the dam will bring water levels up to a height that would completely inundate old towns — including dumps, mines, and factories that will seriously pollute the waters. The Yangtze also affects fisheries — with much less freshwater and sediment getting to the ocean, scientists expect fish catches to decline.

This example shows that even renewable energy sources can have their social and environmental issues. Future hydropower developers need to closely consider the whole picture before they implement large structural changes.

Hydropower plants can play a huge part in reducing fossil fuel use. Eighty percent of Brazil's electricity comes from hydropower. Consequently, the country's power sector produces four times less carbon dioxide than comparable power sectors in other countries. On another positive note, hydropower is one of the cheapest renewable energy options available today because developers and power providers have already developed and built so much of its technology and infrastructure, which are in wide use.

Ocean power

If you've ever spent time on the sea coast, you know that the tides come in and flow out in a daily cycle, pulled by the moon's gravitational force. You may find yourself scooting your towel up the beach a few feet while the hours of the afternoon go by. Not only can the tides move you off the beach; they can move turbines, too.

Ocean power functions in basically the same way as hydropower, using the force created by the movement of water. But rather than coming from river flow, this water power comes from the movement of the currents, tides, and waves. Here's how each works:

- **Currents:** Turbines are placed in flow regions of naturally occurring strong currents.
- **Tides:** At full tide, water is held back with gates. When the ocean reaches low tide, the gates are lifted and the water flows out forcefully, spinning turbines to generate electricity.
- **Waves:** Turbines are put in the areas of strong wave action, and each wave that hits the turbines spins them.

As with almost all renewable energy, tidal power sources have a high start-up cost, but the environmental benefits could be huge; tidal and ocean energy give off no emissions. For countries that have long coastlines, including Canada, the U.S., and Australia, ocean power holds huge potential. France and China are already using tidal power.

Tidal technologies are still being perfected, however. Ocean power developers are currently working on new pilot projects that use more efficient technology that does not involve damming bays or estuaries. These new projects employ turbines on the floor of coastal zones (previous projects worked on the surface of the waves).

From plants to energy

Any herbivore or vegetarian can tell you that plants are full of energy, but some plants can power more than just people and animals. When you ferment plants that are high in sugar, such as corn and sugarcane, they form a kind of alcohol, known as *ethanol,* that you can use as a fuel (or *bio-fuel,* meaning it comes from living organisms).

Hypothetically, scientists consider bio-fuels to be zero-emission. Even though engines emit greenhouse gas when they burn the fuel, that process works in a closed-loop cycle: The carbon dioxide going into the atmosphere is the same carbon dioxide that the plant absorbed from the atmosphere when it was growing. Bio-fuels also release less carbon dioxide when burned than conventional gasoline. Those bio-fuels that are not truly zero-emission are those that use a lot of fossil fuels in the growing process. That is why corn ethanol is less efficient than sugarcane.

Consternation over corn

Ethanol from corn is controversial because it uses a global staple food for fuel. Corn-based ethanol distorts the market because some farmers plant less of one crop (wheat, for example) to benefit from government subsidies for growing corn or grain for fuel. These subsidies are a real factor in the U.S. and Canada. (We talk about government subsidies in Chapter 10.)

Almost everything people eat nowadays connects to corn somehow. You can break corn down into many forms, including corn flour, corn oil, and corn syrup. You can find it far beyond breakfast cereals — you munch on corn when you eat licorice, table syrup, ketchup, and beer. Behind the meat counter, the beef, pork, chicken, turkey, and even fish that you buy eat corn feed. Even your chicken nuggets are about 75-percent corn. Forget food — you can find corn products in your toothpaste and lipstick, and even in your drywall, cleaners, and paper products. . . . Need we say more?

The majority of people in the world — those living in developing countries — depend on corn as the basis of their diet. Consequently, energy experts agree that humanity shouldn't make fuels from food crops, especially corn.

The developers behind corn ethanol say that they won't use corn in the future after they find new sources for ethanol. They assert that corn is like a practice run, aiding them to determine which plant sources work most efficiently in developing fuel. The corn era of ethanol is comparable to the black-and-white era of television — with a technological breakthrough just around the corner.

Other sources for bio-fuel

In Brazil, they use sugarcane as the source of their ethanol production. Sugarcane is a much more effective source than corn, particularly because it produces about seven crops before you need to replant it. For this reason, you need to use less energy to produce it than you do to produce corn. Unlike when you grow corn, you don't need tractors that run on gas or pesticides to grow sugarcane. Sugarcane just grows. Another advantage of using sugarcane to produce ethanol is that sugarcane isn't a staple food crop, so using it for ethanol doesn't risk raising food prices.

Brazil's sugarcane farming is controversial, too, because farmers could clear rainforests to produce sugarcane. The removal of forests contributes 25 to 30 percent of the world's greenhouse gases. (We consider this conundrum in Chapter 12.)

Another possible solution to the bio-fuel problem is to forgo fresh plants entirely. In the following section, we look at how agricultural waste can make ethanol.

Nothing wasted

One person's garbage is another person's alternative energy source. Although our civilization produces a great deal of waste, it can turn some of that waste into fuel for energy. Solid waste (such as plant waste and animal waste, which would otherwise go to the compost or landfill), liquid waste (such as used frying oil), and gases (which our landfill sites emit) all offer power possibilities.

Not your regular smokehouse

In some situations, using biomass as a renewable resource does more harm than good. Two and a half billion people in the world today use biomass — animal poop, farm crop waste, charcoal, and wood — as fuel for daily cooking and heating needs. In many developing countries, people use fires to cook and to heat their small homes. This smoke creates major health problems for the children living in these homes.

International development organizations are currently working to promote solar cooking stoves as a clean alternative, which provides the added benefit of healthier lungs.

Solid waste: Biomass

We can burn the plant and animal waste that currently packs landfills to produce energy — this material is called *biomass.* (Living organic matter burned for fuel is also considered biomass.) Humans have used the simplest form of biomass for thousands of years — wood for fires. People still use wood in fireplaces and woodstoves to this day, and many countries in the developing world depend on wood for both cooking and heating their homes.

Biomass has evolved beyond wood and fire, however. Today, possible biomass sources include solid plant waste from

- **Farms:** Corn stalks, straw, manure
- **Forestry and paper industry:** Bark, wood scraps, sawdust, pulp, wood chips
- **Home:** Kitchen food scraps, yard and garden clippings, sewage sludge
- **Vineyards:** Grape waste after the grapes are processed and crushed

Burning garbage is different than burning biomass. Toxic chemicals released from burning garbage can be hazardous to human health.

Burning biomass emits carbon dioxide (and some nitrogen oxide and sulfur dioxide) when burned. Nevertheless, burning biomass releases fewer emissions than burning coal does.

Burning waste is carbon-neutral only if it comes from plant materials because the carbon dioxide released is only what the plants absorbed when growing.

You can burn biomass along with coal. In this scenario, biomass replaces some — but not all — of the coal burned for energy, lowering the emissions produced. This technology is called *co-firing.* Coal plants can begin co-firing right away because you generally burn both coal and biomass by using boilers that heat water to create steam, which then turns turbines to create electricity. Usually, a coal plant can't replace more than 15 percent of its coal with biomass without losing efficiency. Someone would have to develop a new system designed for biomass to make a plant work efficiently using a higher percentage of biomass.

You can even use biomass right in your own home, never mind what the power plant is doing down the road. Zoë's uncle was the first person she knew to install a wood-pellet-burning stove in his home. The wood pellets are made of wood scraps and burn much more efficiently than logs in a wood stove. Straw pellets are also extremely efficient heat sources. We talk more about energy-efficient changes you can make in your own home in Chapter 18.

You can also turn biomass into bio-fuel, which we talk about in the following section.

Liquid waste

People use renewable sources to create bio-fuel, and bio-fuel can replace petroleum-based fuel in gas and diesel engines. The most common and developed types of bio-fuel are ethanol (technically called bioethanol) and biodiesel. The section "From plants to energy," earlier in this chapter, discusses bio-fuel and ethanol in detail; in the following sections, we discuss how you can use waste products for these fuels.

Ethanol

Ethanol is alcohol based and can be derived from many different kinds of plant material — even plant waste. The plant only needs to contain sugars. This group includes corn, wheat, rice, sugarcane, sugar beets, yard clippings, and potato skins.

You can use straw and switch grass, wood chips, corn husks, and poplar trees as bio-fuel fodder, too. Because these plants aren't high in sugar, they must undergo a special process involving an enzyme that digests their cellulose and turns it into a sugar. This kind of ethanol is called *cellulosic.* Companies such as Iogen and Shell are now making it commercially available.

Cellulosic ethanol does carry a higher cost than corn and grain ethanol due to an additional stage required in processing, but it has considerable benefits. It makes use of agricultural waste, instead of using the crop itself (which can constrain food supplies). And because it comes from existing waste or a naturally growing source, like switch grass, rather than a cultivated agro-business crop, it requires far less energy to produce, resulting in fewer emissions.

Nothing in nature is really waste, so people need to consider the nutrient value of corn husks and straw going back to the soil versus using it as a biomass for fuel.

Biodiesel

Biodiesel is oil based, and people can make it from sources such as used frying oil. Aside from deriving energy from a waste product, you also get the extra benefit of smelling French fries from the tailpipe of any passing car. In Owen Sound, Ontario, a local biodiesel manufacturer uses stale-dated margarine to make a very good fuel. It sells for ten cents less a liter than regular diesel. (I can't believe it's not diesel!)

Gas from garbage

Organic materials are composed largely of *hydrocarbons,* which are made up of hydrogen, oxygen, and carbon atoms. When organic material decays, it releases gases, mainly carbon dioxide (carbon and oxygen) and methane (carbon and hydrogen), made from these atoms. Although it's a potent greenhouse gas (see Chapter 2), people such as farmers outside of Ottawa, Canada, are using methane as an energy source, processing their manure and food waste to power a generator; the waste becomes neutral (and a high-grade fertilizer) and doesn't poison the water supply. You can also capture methane from landfills, major composting facilities, or sewage treatment plants, and then burn it to produce energy. Methane does release carbon dioxide, but at a very low level, when you burn it.

Capturing methane gas from a landfill in Idaho powers 24,000 local homes. The U.S. Environmental Protection Agency's Landfill Methane Outreach Program supports this project and hundreds of others.

Exploring Another Non-Renewable Energy Source: Nuclear Power

Nuclear technology produces electricity around the world. It's also a hot topic of debate. Now that people recognize the climate change in the air, nuclear power has regained favor among some as a low-emission energy source. The International Energy Agency expects the use of nuclear energy to increase in some jurisdictions, but it also expects the overall share of nuclear energy versus all energy sources to actually decrease in the future.

Understanding nuclear power

Nuclear reactors produce electricity through *nuclear fission* (or splitting the atom). Current conventional reactors use uranium, a naturally occurring radioactive mineral, as the fuel for the chain reaction of nuclear fission, when one molecule blasts off an electron to split another molecule. The heat generated by the chain reaction boils water to create steam, which turns turbines to make electricity. Essentially, nuclear power is just a tea kettle on a very dangerous nuclear fire.

Nuclear power is a non-renewable resource because the Earth has a finite supply of uranium, although it has more uranium than fossil fuels. Uranium has its definite drawbacks as a material, which we discuss in the section "Weighing the negatives," later in this chapter.

Looking at the positives

Nuclear energy's supporters cite the following benefits:

- **Less scarce than oil:** Although the Earth doesn't have an unlimited supply of uranium, compared to fossil fuels, uranium is a relatively plentiful resource, available around the world.
- **Low greenhouse gas emissions:** Nuclear power plants produce only indirect emissions, relating to mining and transporting the uranium, building the plant, and (depending on the type of reactor) enriching the uranium.
- **Mature technology:** Unlike the other energy sources we discuss in this chapter, the infrastructure and systems to support nuclear power plants and related mining activities are already in place in some countries.
- **Steady cost:** The price of nuclear energy does not fluctuate as much as the price of energy generated from fossil fuels. (The cost of constructing nuclear generators, however, continues to rise.)

Weighing the negatives

Detractors of nuclear power voice major concerns, including the following:

- **Health concerns:** Long-term health studies vary, but a number of recent studies demonstrate higher cancer rates among populations that live near nuclear reactors. Uranium mining also endangers the health of miners.
- **High capital and maintenance costs:** Nuclear plants are expensive to start up and maintain.
- **Non-renewable resource:** Uranium is finite; when it's gone, it's gone.
- **Reliability:** Nations around the world have varying success with reactor reliability. Some countries, such as Canada, have had persistent problems keeping reactors on line. Breakdowns and retrofits have cost taxpayers billions of dollars.

- **Risk of proliferation of nuclear weapons:** People can use nuclear fuels in nuclear weapons. India made its first nuclear weapon by using spent fuel from a Canadian reactor.
- **Safe storage of nuclear waste:** Nuclear waste must be kept out of the biosphere for at least a quarter of a billion years before it's no longer toxic. No current technology can contain nuclear waste for that length of time.
- **Safety of nuclear plants themselves:** Many people equate nuclear power with the Chernobyl disaster, the near-meltdown of a nuclear power plant in the Soviet Union in 1986. Fears exist that a major accident could happen again. Nuclear reactors do run an extremely small risk of experiencing a catastrophic accident.
- **Security concerns:** Some people worry that terrorists could target nuclear plants and materials because of the great and long-lasting damage such an attack could cause.

Negotiations for what technologies to accept under the Clean Development Mechanism (CDM — which is part of the Kyoto Protocol) ruled out nuclear technologies. (Refer to Chapter 11 for more about the Kyoto Protocol and see Chapter 12 to explore the CDM.) The member countries to the Kyoto Protocol officially decided that nuclear power isn't clean enough for the Clean Development Mechanism. The European Union voiced a similar view of nuclear energy. Working toward a goal of getting 20 percent of energy from renewable sources by 2020, Germany, Spain, and Sweden are committed to shutting down their nuclear plants. France relies on nuclear energy in a big way, however, and so does Switzerland.

Chapter 14

Show Me the Money: Business and Industrial Solutions

In This Chapter

- Making environmental progress in manufacturing
- Trading carbon between businesses
- Improving buildings to reduce greenhouse gas emissions
- Celebrating corporate successes
- Getting help from bankers, insurers, and lawyers
- Growing green with farming and forestry

When it comes to fighting climate change, one thing you may hear businesses tell you is that they can't afford to reduce greenhouse gases and switch to sustainable energy. If people expect companies to spend a fortune on reducing greenhouse gases, business reps say, those businesses will be hobbled in today's competitive marketplace.

In fact, many businesses are already on the greenhouse gas–reduction bandwagon. General Electric is the largest company in the United States. Its CEO, Jeff Immelt, has said that he'll double General Electric's investments in energy and environmental technologies to exploit what he sees as a huge global market for products that help other companies — and countries such as China and India — reduce their greenhouse gas (GHG) emissions. BP, the former British Petroleum (which now goes only by its initials), talks openly today of going "Beyond Petroleum." It used to be an oil company, but it now sees itself as being in the energy business, developing new sustainable alternative energy sources, such as hydrogen and bio-fuels, and working to reduce GHG emissions in other ways. Both companies are concerned, responsible, corporate citizens; also, they understand that when it comes to climate change, corporate responsibility is good for both the planet and the bottom line. They know, too, that if they don't take the lead, the government may — something we talk about in Chapter 10.

Companies can improve how they manufacture products, using modern energy-efficient equipment and recycling. But more than just manufacturers and oil companies can get in on the game. Companies can also get involved in the creation of new green services, or they can get involved in the carbon market. They can change how they construct buildings or turn wood into paper. And you can help them by demanding new, greener products that don't produce as much greenhouse gas, and by rewarding those companies that put sound greenhouse gas–fighting practices in place. You're their customer — and the customer is always right.

Processing and Manufacturing Efficiently

Most manufacturing requires a great deal of energy, usually from fossil fuels that create a lot of carbon dioxide emissions. In many cases, much of this energy is actually wasted, thanks to old and inefficient equipment and weak regulations governing its use. The actual manufacturing process creates even more greenhouse gases.

Taking steps to conserve energy

Manufacturing doesn't have to be wasteful; with some tweaks, industry can use less power, causing fewer greenhouse gas emissions. There are numerous environmental consulting firms all over the world, as well as non-governmental organizations, that specialize in working with businesses, companies, and industry to reduce their greenhouse gas emissions by lowering energy use.

Here are a few steps that the Intergovernmental Panel on Climate Change (IPCC) recommends manufacturers take to help conserve energy:

- **Measure how much energy the manufacturing process uses and how many emissions it creates.** Use this information to set benchmarks and goals for reduction. The industry can measure its success from the changes it makes.
- **Use the correct-size equipment properly and conservatively.** Keep the equipment tuned up and fix malfunctions when they occur. Also, choose equipment such as the optimal size of piping to cut energy use.

You'd be surprised how often mismatched pipes and over-sized motors waste large amounts of energy. Energy guru Amory Lovins, of the Rocky Mountain Institute, estimates that the U.S. could cut electricity use by 40 percent simply by replacing the wrong-sized electrical motors with motors that are the correct size!

- **Use more energy-efficient motors and equipment.** Running motors accounts for over 60 percent of electricity use in European and U.S. industries, and businesses can run motors more effectively by changing materials and improving aerodynamics. Businesses need to improve the efficiency of all pieces of equipment — even fans and pumps — along the way.
- **Insulate buildings and equipment sufficiently.** Insulation keeps any building's energy use low, whether you're trying to keep the heat in or out. Likewise, insulating hot water pipes reduces the loss of heat and also lessens the energy needed to heat the water and compensate for the lost heat.
- **Reduce leaks of any sort (such as air and steam).** For example, when the pressure of the steam drives a turbine, escaping steam reduces efficiency. The boiler has to run that much harder to make up for lost steam. Air leaking into boilers and furnaces can have the same impact. Leaks mean that energy is being spent driving the air or steam into places it shouldn't be.
- **Recycle materials.** Both the steel and aluminum industries have found recycling to be a major advantage. Recycling the steel from old furnaces, for example, makes up a whole third of global steel production and uses 30 to 40 percent of what the process takes if started from raw materials.

Not only can these steps help companies cut back on their greenhouse gas emissions through lower energy use; they can also save companies money by no longer wasting pricey power.

Using energy efficiently

With the money saved through energy conservation, companies can adopt new, efficient technologies for applications such as electric equipment, heaters, and boiler systems:

- **Systems powered by sustainable energy:** Industries can use their own biomass waste, such as wood, food, pulp, and paper scraps, as fuel. Some industries can power themselves by using methane from landfills to run boilers. Solar and wind power are other renewable options. (Refer to Chapter 13 for more about sustainable energy.)

Not every industry can turn to sustainable energy; some manufacturers require a particular fuel, such as the iron industry, which uses coke. (Refer to Chapter 5 for more about the steel industry.)

- **Combining heat and power:** Businesses can use up to 90 percent of the excess heat given off by power production (or generated by machinery) to replace regular heating within a building or buildings, instead of simply pumping that heat out of the building. Businesses in Germany and the Netherlands use this technology, known as *co-generation.*

Table 14-1 shows how much nations can reduce their emissions if their industries switch to more efficient technologies. (A metric megaton [Mt] is one million metric tons, and a kiloton [Kt] is one thousand metric tons.)

Table 14-1 Industries' Emission Reductions with Low-Energy Motors

Region	*Emissions Reduced*	*Equivalent*
European Union	100 Mt CO_2 each year	1/6 of a year of the U.K.'s annual GHG emissions
United States	90 Mt CO_2 each year	1/80 of the U.S.'s annual GHG emissions
Africa (food processing plants, oil refineries, utility companies)	100 Kt CO_2 each year	1/30 of Madagascar's annual GHG emissions

Sources: IPCC, Working Group III: Mitigation, Chapter 7: Industry, p.16; IPCC, National greenhouse gas inventory data for the period 1990–2005, Table 4, p.17.

High-efficiency, new technology can save companies money in the long run, reducing their energy consumption considerably, but the technology isn't cheap to obtain. Companies in developing countries, especially, might balk at the expense of high-efficiency equipment, and understandably so — developing countries often simply don't have the budgets and financial support for energy-effective technology.

Subsidies from national governments for energy-efficient practices can help industries make big changes in how they operate. The Kyoto Protocol offers two programs to help developing countries obtain these new technologies: The Clean Development Mechanism and joint implementation.

- **The Clean Development Mechanism (CDM):** Under this program, industrialized countries pay for clean energy projects in developing countries.
- **Joint implementation:** Through this program, industrialized countries and developing economies partner to implement projects such as capturing methane from landfills and using it to produce energy, or shifting from coal to renewable energy sources.

Both programs help industries in the developing world introduce efficient technologies and reduce greenhouse gas emissions. (We cover the Kyoto Protocol in Chapter 11 and discuss these programs in Chapter 12.)

Considering individual industries

Although the steps laid out by the IPCC (which we discuss in "Taking steps to conserve energy") are relevant to most manufacturers, specific industries face particular challenges.

Cement production, for instance, is a particularly carbon-intensive industry, making up a whopping 5 percent of all carbon dioxide emissions in the world. Parts of these processes are unavoidable, such as the carbon dioxide that the limestone of the cement naturally gives off when the cement forms. However, the industry could benefit from using energy from renewable resources or from systems that capture and store the carbon dioxide emissions underground. (We talk about storing carbon dioxide in Chapter 13.)

Other greenhouse gas–intensive industries have found some innovative ways to reduce their emissions. Here are a couple:

- **Pulp and paper:** Canada's forestry giant Tembec has managed to reduce its production of greenhouse gases directly and indirectly. It recycles wood chips and other waste as fuel, burning them in place of higher carbon-emitting fossil fuels. (This kind of fuel use is sometimes called a *closed loop system,* in which you don't input to or output from the system — it's a full cycle.) Tembec now dries its pulp by using more efficient hot air dryers, replacing earlier steam units that were fossil-fuel powered. Tembec even extracts sugar from the *cooking liquor,* the fluid left over after paper is manufactured, and turns it into ethanol, which it then sells as a product to be used in things like antiseptics and sanitizers. Flip over to Chapter 13 for more on using plants and waste as fuel.

  Thanks to steps such as the ones Tembec has taken, the Forest Products Association of Canada boasts that it has met and far exceeded the Kyoto target of reducing carbon dioxide emissions to 6 percent below 1990 levels. They've actually reduced their emissions to 42 percent below 1990 levels!

- **Aluminum:** Many aluminum companies are concentrating today on recycling. To recycle aluminum, these companies need to use only 5 percent of the energy they would need to manufacture it from raw ore. One manufacturer, Alcoa, plans to boost the percentage of recycled aluminum it uses in new production to 50 percent by 2050.

Thanks to the introduction of newer technology, the aluminum industry has been able to reduce its production of perfluorocarbons, the particularly nasty greenhouse gas that captures from 6,500 to 9,200 times more heat than carbon dioxide.

Trading Carbon between Manufacturers

Some manufacturers and producers can reduce their greenhouse gas emissions more easily than others. To address that imbalance and enable industries to reduce their greenhouse gas emissions across the board, some jurisdictions and commodity traders have created carbon markets. The *carbon market* isn't like a flea market, the sort of place you drop by on a Saturday morning to pick up a bargain in chunks of coal.

Here's how a private carbon market works:

1. Companies form a group and make a commitment to each other.
2. They agree on how they want to reduce their emissions over the year individually and collectively.
3. If the company reduces its emissions more than planned, it has *carbon credits,* which it can sell. If a company doesn't make its goal, it can buy someone else's credits. (Another name for this process is *cap and trade.*)

Companies can actually make money by reducing their output, creating carbon credits and selling them. The carbon market ensures that carbon dioxide levels are being reduced. It just doesn't worry about where or by whom.

Although the majority of carbon markets around the world are government initiatives, businesses can and have implemented emissions trading themselves. (We discuss government-led carbon markets in Chapter 10.) The Chicago Climate Exchange is the world's first voluntary carbon market. It goes beyond carbon dioxide to include almost all greenhouse gas emissions. The organizations involved in this exchange make a voluntary, yet legally binding, commitment to reducing emissions. Participants range from universities to retail businesses to power-generating companies — Rolls Royce, DuPont, and Sony Electronics are among the members that have committed to reductions. Internal committees regulate the system, and to make sure everyone's living up to their commitments, a third party oversees the emission reductions.

Adaptation for industry

Industry is as vulnerable to the effects of climate change as are communities and individuals. Businesses need to adapt to these changes, just like everyone else.

In part, they need to make mundane, obvious adaptations — renovating facilities to anticipate extreme weather, for example.

Mining companies are re-engineering *tailing ponds,* the holding ponds for mine waste, to prepare for more extreme weather events. Diamond mines in the Arctic are adapting to the warming weather — the winter roads, relying on consistently frozen ice and snow, on which they used to rely are now available for less of the year, due to climate change. This is driving up their costs because they have to fly more goods in.

Industry can also make proactive adaptations, however, involving diversifying and thinking in new ways. BP is a prime example of smart adaptation. By broadening its mandate beyond oil, it can supply the market's energy needs, whether the world depends on oil, ethanol, solar power, or wind energy.

Any major production or manufacturing company can take similar actions. Forestry and agriculture-based companies, for example, can diversify their resource base so that they harvest from numerous small locations, rather than one large location. This adaptation can allow them to continue supplying wood to the market, even if climate change and its effects impact one area, for example, with forest fires or droughts.

Beyond do-gooding, businesses can make money in carbon trading. Goldman Sachs, a U.S. investment bank, owns shares in the Chicago Climate Exchange and the European Climate Exchange. (We talk more about how banks are getting involved in fighting climate change — and profiting, in the process — in the section "Support from the Professional Service Sector," later in this chapter.)

Building Greener Buildings

Production and services aren't the only ways that companies can fight the good greenhouse gas fight. They can make a difference by changing how they do business, but also where they do it — their offices and factories. These big buildings currently emit a lot of greenhouse gas, but that means they also present a big opportunity. In fact, the IPCC reports that improving the efficiency of commercial and industrial buildings is a more cost-effective way of reducing greenhouse gases than overhauling industry's manufacturing processes.

Cutting back on heating and cooling

Businesses often use the most energy keeping an office or factory warm in the winter or cool in the summer. Those businesses can cut back on energy consumption by installing heating and cooling systems that don't guzzle so much (or any) carbon-emitting fuels. In Chapter 18, we look at different heating and cooling options for houses, and these solutions also apply to companies. Companies can reduce their heating- and cooling-related energy consumption in some very simple and inexpensive ways:

- **Insulation:** A properly insulated building doesn't require nearly as much energy to heat or cool. You lose about 60 percent of the heat pumped into a poorly insulated building through the roof and the walls alone.
- **Air circulation:** Buildings in moderate climates can benefit from using systems that either keep the outside air out or let it in, depending on whether you need to heat or cool the building on that day.
- **A green roof:** A rooftop that features layers of grass, soil, and water-proof lining naturally cools a building, providing insulation and reflecting the sun's light (unlike black asphalt roofing, which absorbs it). In fact, a green roof reduces air-conditioning demand by 25 percent. The plants on these roofs have the added bonus of absorbing some extra carbon from the atmosphere.
- **A white roof:** For roofs that can't support plant life, a simple lick of white paint can help cool buildings in hot climates because the white paint reflects light.

Good insulation coupled with good natural lighting can even heat a whole building. The Rocky Mountain Institute in Colorado is (as the name suggests) located in the Rockies and is under heavy snow much of the winter, yet the building has no furnace. It relies instead on the sun's rays, which stream in through giant windows, and it holds that heat through maximum insulation. They even grow bananas indoors to show the building's balmy conditions. Check out founder Amory Lovins's Web site for more wonderful information on the enormous potential of energy efficiency: `www.rmi.org`.

Exploring energy alternatives

Business and industry can lead the charge in shifting buildings away from consuming fossil fuels and toward renewable resources. Because their buildings consume so much energy, when they make a change, it has a big effect. Wal-Mart launched a renewable energy initiative in 2007, committing to using solar power for 30 percent of the energy needs at 22 of its stores — a tiny fraction of the almost 6,500 Wal-Mart stores worldwide. This seemingly small initiative ranks among the ten largest solar installations in the world because of Wal-Mart's multinational reach and number of buildings it controls. Wal-Mart's ultimate goal is to become 100-percent reliant on renewable fuels and produce no waste. (For an overview of the energy alternatives that businesses can investigate, check out Chapter 13.)

By creating its own energy from renewable sources, a building can remove itself from the electricity grid — or, at least, reduce its use of energy coming from the grid. The more energy a building can produce on its own, the more money the owner saves. In some jurisdictions, building owners can even sell their excess renewable energy back to the grid and make a profit.

Certifying new buildings

Improving efficiency and energy sources can cut back on the greenhouse gas emissions of existing business buildings, but when companies require new buildings, they have the opportunity to really go green. They can ensure that their new digs are as greenhouse gas–smart as possible by following environmental standards. Having a standard for buildings sets the bar for companies. And while companies improve and compete, they raise the bar by raising standards.

In North America, LEED (Leadership in Energy and Environmental Design) leads the way. The United States Green Building Council created and constantly modifies the LEED to reflect new practices and materials. The LEED provides a set of standards that new or renovated buildings must achieve to be certified. LEED standards encompass all aspects of a building's construction, including the following:

- The percentage of demolition material recycled (in the instances when a building was demolished to make way for new construction)
- How efficiently the building uses water
- The materials used and what percentage of those materials came from less than 500 miles away

LEED also has a certification program for architects and engineers to confirm their understanding of green building products and practices. To find out more about LEED in the United States, see www.usgbc.org. For LEED in Canada, check out www.cagbc.org.

Other countries have programs similar to LEED:

- **United Kingdom:** The Building Research Establishment and Environmental Assessment Method (BREEAM) is the first set of environmental building standards established anywhere — developed even before LEED. BREEAM has specific requirements for the widest range of buildings — from schools to warehouses to theaters. They even have a ranking for prisons! (You can visit the BREEAM site at `www.breeam.org`.)
- **Australia:** Green Star is run by the Green Building Council of Australia. Green Star was inspired by BREEAM and LEED, but modified to better fit the Australian environment. Its particular focus is office buildings, from design, to construction, to retrofits. Green Star hands out stars (up to six) to buildings that meet its environmental standards. It plans to expand its standards to cover industrial, retail, and residential buildings. (For more info, visit www.gbca.org.au.)

Australian toolmaker Bordo International's new head office was awarded five stars; their project design is expected to reel in a 68-percent energy savings, thanks to effective insulation, windows, and blinds.

Corporate Success Stories

If you zone out whenever you hear the word "corporate" because it seems so big and heartless (or just plain boring), we want to change that. Corporations are using their big profiles (and big budgets) to help reduce emissions. Table 14-2 profiles some of the corporations that are ahead of the pack in reducing their carbon footprints.

Table 14-2 Businesses Reducing GHG Emissions

Company	*Target*	*Successes*	*Savings*
Barclays (banking)	–20% GHG emissions by 2010 (versus 2000)	Carbon neutral and running off 50% renewable energy	Yet to be recorded
DuPont (chemical manufacturer)	–65% GHG emissions by 2010 (versus 1990) and using 10% renewable energy	–67% GHG emissions, +35% in production, and +3% in renewable energy	$2 billion saved, plus $10–$15 million annually

Intel (computer manufacturer)	–10% PFC* emissions by 2010 (versus 1995) and –4% energy use by 2010 (versus 2002)	–35% PFC emissions and –12% energy use	$10 million annually
Johnson & Johnson (health care products)	–7% GHG emissions by 2010 (versus 1990) and increasing renewable energy use	–17% CO_2 emissions and +39% of electricity from renewable energy	+372% in sales and $30 million saved annually
Wal-Mart (discount store)	–30% energy use, –20% GHG emissions in 7 years, and use of 100% renewable energy	$500 million annually invested in energy-efficient technologies	Yet to be recorded
Toyota (car manufacturer)	To become an environmental leader	–2.5% CO_2 emissions in 2 years and –35% in energy use in 5 years	Yet to be recorded
Interface (floor-covering manufacturer)	Zero environmental impact by 2020	–45% energy use, –60% GHG emissions (versus 1996), and +16% use of renewable energy	$330 million from waste reduction
Eastman Kodak (photography company)	–15% energy use and –20% CO_2 emissions by 2003 (versus1997)	–12% energy use and –17% CO_2 emissions by 2005	$10 million annually
HDR, Inc. (architecture and engineering)	Use sustainable development principles in all projects	Spent $2 billion in renewable energy projects and built 5.8 million sq ft of LEED** buildings	$40 million saved annually in client energy costs

** Perfluorocarbon (PFC), a greenhouse gas.*

***Leaders in Energy Efficiency and Design (LEED).*

Source: The Climate Group, Case Studies, Corporate `http://theclimategroup.org.`

Support from the Professional Service Sector

Although some people might say that bankers, insurers, and lawyers produce a lot of hot air, their businesses don't immediately come to mind when you think of global warming. Nevertheless, companies in those sectors have seen the silver lining of profitability in the climate change cloud and have gotten involved with ventures that help others cut back on carbon.

Banking on the environment

Many banks are becoming involved with the fight against climate change by offering specialized services to clients committed to reducing greenhouse gases or providing renewable resources:

- **Goldman Sachs:** American investment bank Goldman Sachs has invested heavily in carbon markets, owning large shares in the Chicago Climate Exchange and the European Climate Exchange. (We talk about private carbon trading in the section "Trading Carbon between Manufacturers," earlier in this chapter.) Goldman Sachs also founded the Center for Environmental Markets, which issues grants for research in market solutions for environmental issues. The company has invested over $1.5 billion in renewable energy sources.
- **Bank of America:** This second-largest U.S. bank has created an environmental banking group dedicated to conservation and reducing global warming. The bank's new headquarters in New York is housed in one of the greenest skyscrapers ever built.
- **ABN AMRO:** This Dutch banking giant, which refers to climate change on its Web site (`www.abnamro.com`) as "both a challenge and an opportunity," offers risk management services to help its clients reduce their possible losses from climate change.

Insuring against climate change

The insurance industry has a vested interest in stopping climate change; the extreme weather it will bring (which we discuss in Chapter 7) will result in a huge surge in claims. Knowing that investing today can prevent giant payouts in the future, American International Group (AIG) offers financial support to projects that encourage greenhouse gas emission reductions. AIG says it may invest in forests, renewable energy resources, greenhouse gas mitigating technologies, and green real estate. The Zurich-based Swiss insurance company Swiss RE is also interested in working to reduce and profit from the danger of climate change, and in 2008, it was named "one of the world's 100 most sustainable companies" by Innovest Strategic Investment Advisors.

The insurance industry is actively funding research efforts. In Canada, insurance companies fund the Centre for Catastrophic Loss Reduction at the University of Western Ontario. In Bonn, Germany, the Munich Climate Insurance Initiative (MCII) is helping to develop alternative insurance products that can facilitate both spreading climate-related risks (ensuring that no one insurance company shoulders the burden of paying out for the aftermath of extreme weather events) and adaptation-response measures. Members of the MCII include the International Institute for Applied Systems Analysis, German Watch, the Potsdam Institute for Climate Impact Research, and individuals from the World Bank and Munich Reinsurance Company. The insurance industry largely funds the effort.

Making it legal

Many law firms are actively engaged in making the fight against global warming a legal imperative. Firms such as the U.S.-based Baker & McKenzie offer to help government clients worldwide develop climate change laws and regulations. Firms also offer services on carbon markets, carbon-offset projects, and trading emissions. The legal framework for such projects, even those outside government regulations, requires careful drafting.

But beyond the services they offer, law firms are businesses like any other. They have an impact through the energy they use and the paper waste that they produce when they work with clients on important issues. The U.S. Environmental Protection Agency and the American Bar Association have partnered up to address climate change issues through a voluntary program. They created the Law Office Climate Challenge (`www.abanet.org/environ/climatechallenge`). The program has over 60 law firms signed on to one of three commitments:

- Buying renewable energy
- Lowering the amount of waste produced
- Cutting office energy use by at least 10 percent

All the actions in this simple and effective program can help reduce the law profession's carbon footprint.

Farming and Forestry

Farming and forestry are uniquely posed to make a difference in the fight to stop global warming. Like all industries, they can cut back on their greenhouse gas emissions by improving their energy efficiency and moving to sustainable energy sources. But what makes them truly exceptional is that they can actually increase how much carbon dioxide is absorbed from the natural greenery and soils under their management. Talk about a global warming one-two punch! How agriculture and forestry use the land has created a third of global greenhouse gas emissions — which means that these industries can become a huge part of the solution to climate change.

Supplying bio-fuels

The waste that forestry and farming produces, such as wood, crop waste, and manure, doesn't have to be wasted (although leaving waste in the forest or the fields isn't necessarily wasteful — this material can provide nutrition to the land). All that stuff is actually *biomass,* biological material that humans can use, either by burning it to create energy or turning it into *bio-fuel,* which transportation devices can use. Bio-fuel gives off fewer greenhouse gas emissions than fossil fuels when burned. People are increasingly using bio-fuel as either an alternative to diesel fuel or as an additive to it, which creates a lower-emission fuel. (You can read up on bio-fuel in Chapter 13.)

Beyond providing the materials for bio-fuels, agricultural and forestry practices can benefit from using the bio-fuel themselves. Although companies would have to make an initial minimal investment to convert diesel engines to run on bio-fuel, this investment would likely offer long-term savings, particularly while oil costs continue to rise.

Improving land management

Land management includes many elements. For forestry, land management involves how and where companies grow the trees and what kind of harvesting methods they use. For agriculture, it involves how farmers till the soil, what they add to the soil, and how they grow and harvest the crops. Land managers can engineer all aspects of their operations to be more environmentally friendly, sucking extra carbon dioxide out of the air.

Forestry

The number-one, land-management solution recommended by the IPCC for forestry is to decrease the areas deforested.

The IPCC says that forestry practices must change quickly to counteract rapid, worldwide deforestation. The IPCC warns that forestry companies need to know how climate changes will affect their forests. These changes could include an increase in how fast wood decomposes, as well as more intense droughts and forest fires. (Climate change won't be bad news for all forests initially, however. Because of temperature increases, some trees could grow faster and take in more carbon dioxide.)

Instead of deforesting entire areas, the forestry industry needs to adopt more sustainable methods, such as *selective harvesting*. This method involves removing small groups of trees, leaving behind a range of trees of different ages and sizes. Selective harvesting has many benefits:

- **Helps land stability:** The root systems of trees hold the soil together and assist in the prevention of landslides.
- **Keeps the forest functioning as an ecosystem:** This enables it to continue to serve as habitat for wildlife.
- **Keeps the soil healthy and productive:** Healthy soil, supported by trees, takes in rain — a lot of it. Trees also make the soil more drought resistant by shading it and giving it nutrients.

Ecologists encourage selective harvesting in temperate forests primarily to protect wildlife habitat and biodiversity. Temperate forests include the boreal region, a large band of forests, through Russia, Scandinavia, Alaska, and Canada, which makes up about one third of the planet's remaining forests. These forests have deep soils that can support new growth after clear-cutting. The second-growth forest lacks the species diversity of the primary forest, but only very rarely does a clear-cut forest in a temperate region result in true deforestation.

Selective harvesting is essential to the survival of forests in the tropics, however. Tropical forests grow on very thin soils, which are unlikely to be able to support life after the forest canopy has been cut away. Clear-cutting a tropical forest results in true deforestation.

The forestry industry can implement the sustainable practice of lengthening the time between *rotations* (the time between logging the forest, allowing regrowth, and coming back to log it again), allowing a forest to grow for a longer period of time before returning to log that area again. This longer cutting cycle would increase the carbon uptake of the forests; although young forests absorb carbon more quickly than old trees, older trees can retain far more carbon.

Some in the forestry industry are already taking steps, through the Forest Stewardship Council (FSC), to ensure that forests are sustainably managed. If a forest complies with the FSC's Principles of Responsible Forest Management, the FSC certifies that forest, enabling the operators to use the FSC logo.

According to the FSC, more than 95 million hectares have been certified — about 7 percent of the world's industrial forests — and this number is growing rapidly. Large companies dedicated to helping forest management help make these kinds of programs a success — from FSC-certified chairs to eyeliner pencils. (Check out `www.fsc.org` for more about the Forest Stewardship Council.)

Governments and non-governmental organizations are taking steps to reduce deforestation in the tropics, where trees take in carbon dioxide all year round, making deforestation reduction in tropical regions most effective — and most urgently needed. (Boreal trees in the north don't take in carbon dioxide in winter months.) The majority of the world's rainforests are found in developing countries. The Clean Development Mechanism, a program under the Kyoto Protocol, encourages industrialized countries to fund sustainable forestry practices in developing countries that can help those developing countries cut greenhouse gas emissions. (We talk about this program in Chapter 12.)

Farming

Farming isn't an obvious culprit when it comes to carbon emissions, but it's a considerable contributor. Humanity's ever-growing food needs and desires have pushed the farming industry to deforest valuable land and use emission-heavy methods of farming. Fortunately, greener options are possible.

Location, location, location

As we discuss in Chapter 5, farmers are clearing large portions of the Amazon rainforest to make way for more farmland. This deforestation is disastrous for the climate. The most climate-friendly farms are situated on land that doesn't require cutting down rainforest. Meanwhile, both industrialized and developing countries are displacing land for agriculture in favor of development. Paving over land already in use for local agriculture often forces people to clear forests to plant their crops.

The world needs large-scale solutions to address deforestation for farmland. Governments in countries where deforestation for farmland is a major issue, such as Brazil, need to firmly regulate land use so that they can begin to deal with this major climate issue. We discuss how Brazil's dealing with deforestation in Chapter 12.

Dealing in dirty solutions

Sometimes, fighting climate change can be a dirty job. Healthy soil is a critical partner in absorbing carbon, but it can also be a major source of carbon dioxide emissions. If farmers modify farming practices to protect the health of the soil, they could cut back on 89 percent of agriculture's carbon dioxide emissions.

Get a green belt by planting a tree

Nobel Peace Prize–winner Wangari Maathai created the Green Belt Movement, which is actively planting trees across mid-Africa. The problem in Wangari's eyes was simple: Climate change and related environmental problems could be traced to deforestation. The solution was just as clear: Plant trees — a lot of them.

Starting in the 1970s, she engaged women in a grassroots effort that led to planting over 20 million trees. Her Kenya Green Belt Movement has since gone international, drawing in other African countries, including Uganda and Zimbabwe.

Maathai's new goal, which she announced at the UN Climate Change Conference in 2007, is to get Africans to plant 1 billion trees.

You can find out more about the initiative at `http://greenbeltmovement.org`.

The first step that farmers can take to reduce carbon dioxide emissions is to change how they manage weeds so that they don't have to till the soil as often — if at all. In short, the less farmers disturb the soil, the better. No-till agriculture increases carbon sequestration (refer to Chapter 2) in the soil.

Farmers can also cut their greenhouse gas emissions by carefully using nitrogen fertilizer, which adds nitrogen dioxide into the atmosphere. Careful use of this fertilizer means simply figuring out how much nitrogen you need to add so that you don't use more than necessary, which can help ensure that farming gives off fewer emissions.

Farmers can also move away from chemical fertilizers and explore greener alternatives, such as organic farming methods. Producing and transporting chemical fertilizers, which are generally made from fossil fuels, is extremely energy-intensive. Studies have shown that organic farming methods use about half of the energy of conventional farming, and also sequester more carbon dioxide in the soil.

Reducing rice farms' emissions

Currently, land management among rice farmers is particularly poor and results in the emission of methane, a particularly potent greenhouse gas. In flooded rice fields, organic material breaks down, and because the still waters form an *anaerobic* (oxygen-free) environment, that breakdown results in the release of methane. Rice farmers can reduce these emissions by keeping the soil dry in the non-growing season, adding any organic materials the soil needs during that dry period, rather than leaving the fields flooded to supply nutrients. In a dry field, *aerobic decomposition* (decomposition that occurs in an environment with oxygen) can happen, which doesn't produce methane.

The global-warming beef with cows

One of agriculture's major sources of greenhouse gas emissions is livestock. With the global diet changing to include more meat, the world has an ever-increasing number of cows, which means an increase in methane. Cows are really gassy creatures.

Scientists around the world are currently exploring different ways to help cows cut this socially impolite and environmentally unfriendly habit. In the United Kingdom, research is underway to see whether adding garlic to a cow's diet can help cut back on the gas. German scientists are testing a pill that traps and eliminates the gas in a cow's first stomach (they have four). In Australia, scientists are experimenting with transferring bacteria from kangaroos' stomachs to cows'. (Kangaroos have a similar digestive system and diet to cows, but they don't suffer from the same gastric unpleasantness.)

Chapter 15

Activists without Borders: Non-Governmental Organizations

In This Chapter

- Understanding the role of different organizations
- Talkin' 'bout the young generation
- Jumping on board an NGO

When you hear the word *activist,* you might think of angry marchers shouting slogans and hoisting protest signs — and you wouldn't be wrong. But you can be an activist in more than one way. Around the world, people have banded together into groups, determined to prevent a global warming disaster. Some of these groups do indeed hold rallies and stage sit-ins, but others are far more comfortable in boardrooms.

One of the most powerful forces in the world arises from people working together to improve society. Organizations in pursuit of social goals have been around for nearly 200 years, fighting against slavery, campaigning for the right of women to vote, and protecting the natural environment. Today, groups of this kind are called non-governmental organizations, or simply NGOs.

The "non-governmental" part of NGO is critical. Because these groups aren't tied to any government, they can be single-minded, focusing on one thing: for example, fighting global warming. In this chapter, we take a look at these organizations, how they're working hard to realize their goals, and how you can get involved.

Understanding What Non-Governmental Organizations Do

If you enter either *global warming* or *climate change,* and then *non-governmental organization,* into a search engine on the Web, you get thousands of hits, listing countless groups. The sheer quantity of NGOs might seem excessive; after all, aren't they all working for the same goal?

The number of NGOs indicates the enormity of the problem of global warming and just how much work needs to be done. Every NGO represents a different segment of the world's population affected by climate change, and each group tackles the problem differently. Some groups strive to increase awareness about the issues. Some stay on the periphery of society, aggressively trying to provoke people into action, and others work from within both government and industry, attempting to prompt progress.

Educating people

The scientists doing research on global warming aren't writing for a broad audience; they're writing for their peers, to further scientific knowledge. This communication is very important because it ensures that science is constantly moving forward, building on new discoveries. Unfortunately, laypeople — the public at large — don't always hear about important research. That's where NGOs come in.

NGOs take complicated technical scientific reports, which are available to the general public, and translate them into language that the average person can understand. These people working as "translators" between science and the public are very familiar with the science and also understand how to communicate it in non-science terms. Many NGOs, such as the World Wildlife Fund and Greenpeace, write climate change reports that take key pieces from scientific reports and make them digestible to the reader. NGOs communicate via fact sheets, brochures, and Web sites, as well as at conferences and workshops, and even in material prepared for school curricula. These organizations hope that they can present the information in a way that motivates people to act.

As a result, NGOs fret over even the most essential terms to ensure that the words have the biggest possible impact. For instance, NGOs around the world struggled over what to call this human-made climate crisis. Global warming? Climate change? Most groups opted to call the threat "climate change" because some places will actually get colder when the global average temperature increases. U.S. groups went with the more popularized "global warming," a term with more immediate impact.

When translating science to lay language, NGOs need to be careful not to distort the facts. That's why, quite often, NGOs ask scientists to review their work to make sure it remains accurate. (Similarly, some of the leading scientists in the world reviewed this book to make sure we properly presented the science!) Many of the larger international environmental groups even have scientists on staff. For example, an internationally respected scientist from the Potsdam Institute, Dr. Bill Hare, also works for Greenpeace.

Keeping watch

NGOs play a key role as watchdogs over government and industry. They keep a close eye on what impact government and industry are having on climate change, and they're quick to point out when the powers that be don't live up to their green commitments.

These organizations are a big part of the reason that climate change issues make so many stories in the media — journalists often depend on organizations to share the top-hitting climate news story and connect them with the right people to talk to. (See Chapter 16 for more about the media.) Watchdog groups strive to ensure that the general public is as well informed as possible on climate change issues and what industry, government, and businesses are — or aren't — doing.

Organizations often publish "report cards" that score businesses or governments on their action on climate change — a clear way to communicate the state of the success of these actions, which the public might not otherwise know about.

World Wildlife Fund, for example, exposed Nike's high-emission running shoes and raised so much awareness about it that Nike agreed to change the gas used in the shoe air pockets, greatly reducing its greenhouse gas emissions. Sinks Watch, as another example, is a project of the World Rainforest Movement, whose goal is to track and critically assess Kyoto Protocol carbon sink projects. (Check out Chapter 11 for more about the Kyoto Protocol.)

Getting the word out

Some NGOs have a reputation as rabble-rousers, constantly raising a ruckus. Greenpeace, for example, is famous (or is that infamous?) for unfurling enormous banners from high-rise buildings, off suspension bridges, or anywhere else their efforts can draw attention — even if it's against the law.

These groups know that their banners won't make an immediate difference. They're doing it for the guaranteed media coverage of their dramatic actions, which ensures that people will address the issue in a public forum.

Following the money

Although most NGOs are working toward a common goal, they're not all on the same page. Some organizations give voice to the very few scientific skeptics. In those cases, determining where a group's funding comes from can be educational.

The American Association of Petroleum Geologists (AAPG), for instance, gave a "journalism" award to Michael Crichton for *State of Fear,* a science fiction novel that painted global warming as a huge conspiracy. (We take a closer look at this book in Chapter 16.) Not coincidentally, the AAPG's research is often funded by the oil industry. Another group, the American Enterprise Institute (AEI), famously offered $10,000 to scientists to challenge the findings of the Intergovernmental Panel on Climate Change. Oil giant ExxonMobil funds the AEI.

Whether hanging banners or organizing mass protests, these acts are most certainly attention grabbers. But not everyone loves these actions — not even everyone in the climate change awareness community is a fan. Some argue that the time these NGOs spend in these attention-drawing endeavors would be better spent working with industry or government to find solutions, rather than slamming those industries and governments with banners. Advocates for the stunts point out that direct actions bring a visual element to global warming, a threat that looms like a slow motion tsunami which is hard for media to cover.

Working with industry and government

From much of the news coverage you see, you might think that all environmental groups spend their time yelling at corporations and governments across barricades — and that industry and government spend their time trying to build bigger barricades. That's not actually the case (although that tends to be the kind of conflict that makes news). A very large number of NGOs make progress by working with industry and government.

Corporate cooperation

Environmental group representatives are often asked to sit on advisory panels for industry and business. These panels are composed of people who offer companies outside perspectives on their plans and actions. For environmental activists, being on advisory panels presents opportunities to help companies develop a greater understanding of the ecological impact of their actions.

The benefits of NGO–corporate partnerships go beyond advisory panels, however. Businesses are a link for NGOs to create policy changes, and the support of an NGO can help endorse the climate-friendly practices of a business. With their comprehensive understanding of the causes of climate change, NGOs can offer industries real insight into problematic practices that have been causing major emissions. And businesses can often find practical (and economical) ways to implement greenhouse gas-reducing technologies and strategies. By bringing these two specializations (so to speak) together, industries can get a more holistic perspective on best practices for business and industry operations.

Some partnerships between NGOs and businesses, such as the following, have really helped reduce greenhouse gases:

- The Climate Group is an international organization made up of companies and government representatives to advance leadership on climate change, work with partners such as HSBC Holdings and Intel to improve management systems, save energy, set targets for reducing emissions, and make a profit while they're at it.
- Conservation International partnered with Ford Motors to create the Center for Environmental Leadership in Business. The Center is dedicated to bringing in large corporate partners, such as Marriott International (the hotel chain), to reduce greenhouse gas emissions and engage hotel guests in rainforest conservation projects. Another partner is Starbucks, now working on a five-year climate change adaptation project to support coffee-growing countries to protect the agricultural land, water, and forests that sustain the countries.
- The Pembina Institute in Alberta, Canada, has worked with companies such as Suncor in the oil industry to reduce their emissions.
- World Wildlife Fund International leads a project entitled Climate Savers, engaging companies such as IBM and Lafarge to reduce their carbon dioxide emissions. The goal is to deliver emission cuts equal to taking 3 million cars off the road every year by 2010.

An NGO risks its credibility when it puts its name beside that of a business. NGOs are rooted in value systems and the values of their members. The World Wildlife Fund faced criticism when it partnered with Nike in 2001 because of the company's labor policies, even though WWF was linked to Nike on only environmental issues. Likewise, corporations can stand a risk of being linked with strong standpoints that a partner NGO might bring to the media.

Government action

On national and international levels, NGOs are working with governments to fight climate change. In individual countries, many governments recognize and often draw on the expertise of NGOs — despite the fact that such groups often disagree with government policy and are working to change it.

Banding together

From the earliest stages, the non-governmental organization (NGO) community realized the climate crisis would require a global solution. So, the NGOs formed their own global network, the Climate Action Network (CAN), which comprises more than 350 NGOs from over 80 nations. Through CAN, NGOs share information and strategies to push for progress internationally, nationally, and regionally.

You can check out CAN's Web site at www.climatenetwork.org. Individual member countries have Web sites, too, with useful links and resources:

- **Australia:** www.cana.net.au
- **Canada:** www.climateactionnetwork.ca
- **United States:** www.usclimatenetwork.org
- **Western Europe:** www.climnet.org

Governments and non-governmental organizations are naturally complementary to one another. NGOs are generally created to fill a gap that people have identified in government. Both the government and NGOs have a mandate to act in the best interest of the public, but NGOs focus on particular issues. NGOs aim to bring the voice of the public to the government.

The U.K. government and The Climate Group collectively launched the Together campaign. Together is an aptly named campaign for convincing people and businesses to make changes in their lives and operations. The Together campaign is already a third of the way to its goal of saving a metric ton of carbon dioxide for every household in London.

Internationally, large NGOs are very active, participating in every climate negotiation since the first meetings toward a global treaty began in 1990. The United Nations (UN) recognizes credible and well-organized NGOs as observers to global negotiations. The role of *observer* isn't nearly as passive as the name may suggest; observers can speak at meetings, but they can't vote on negotiated text for treaties.

In their speeches at these conferences, NGOs can ensure that the grassroots are represented and that the issues aren't lost in any political maneuvering. They also meet regularly with the delegations of governments to discuss policy and political stances. To bring awareness to the public, NGOs communicate with reporters and greatly help focus the attention of the media on key issues. The collective work of these NGOs helps move the negotiations forward and puts pressure on the politicians at the table.

Meeting This Generation

People under 25 years of age make up almost half of the world's population. That's a big share, and because many of the major climate changes are projected to hit 50 years from now, youth groups have a special role in the climate change NGO world. Youth organizations can make a difference by educating their parents and larger community, by greening university campuses, by pressing governments for change, and by promoting a low-carbon lifestyle.

The voice of a child

Probably no one was as young or as effective in a major United Nations Summit as a 12-year-old girl from Vancouver, Canada, in the 1992 Rio Earth Summit. Severn Cullis Suzuki spoke to the delegates and reminded them that they should let their actions match their words:

Coming up here today, I have no hidden agenda. I am fighting for my future. Losing my future is not like losing an election or a few points on the stock market. I am here to speak for all generations to come. I am here to speak on behalf of the starving children around the world whose cries go unheard. I am here to speak for the countless animals dying across this planet because they have nowhere left to go. I am afraid to go out in the sun now because of the holes in our ozone. I am afraid to breathe the air because I don't know what chemicals are in it. I used to go fishing in Vancouver, my home, with my dad until just a few years ago we found the fish full of cancers. And now we hear of animals and plants going extinct every day — vanishing forever.

I'm only a child, yet I know we are all part of a family, 5 billion strong — in fact, 30 million species strong — and borders and governments will never change that. I'm only a child, yet I know we are all in this together and should act as one single world towards one single goal. In my anger, I am not blind, and in my fear, I am not afraid of telling the world how I feel.

I'm only a child, yet I know if all the money spent on war was spent on finding environmental answers, ending poverty and finding treatises, what a wonderful place this Earth would be!

You are deciding what kind of world we are growing up in. Parents should be able to comfort their children by saying, "Everything's going to be alright," "It's not the end of the world," and "We're doing the best we can." But I don't think you can say that to us anymore. Are we even on your list of priorities? My dad always says, "You are what you do, not what you say." Well, what you do makes me cry at night. You grownups say you love us, but I challenge you, please make your actions reflect your words. Thank you.

From the White House to green initiatives

After leaving the White House, President Bill Clinton's vice president, Al Gore, has become one of the world's most prominent leaders in the fight against climate change — but Clinton has been very active in the field, too. The William J. Clinton Foundation, which has many relationships in Africa because of the work it does to combat the spread of HIV-AIDS, is increasingly moving to climate change work. The Clinton Foundation founded the Clinton Climate Initiative in 2006. The project aims to make large-scale energy reductions in order to cut greenhouse gas emissions.

The first phase of the business-based project is to work with 40 cities around the world in an initiative called C40, or the Large Cities Climate Leadership Group. The Clinton Climate Initiative is helping these cities reduce greenhouse gas emissions by creating tools, programs, and products that these cities can use. Local government, companies, building owners, and banks are all part of the game plan, retrofitting 300 municipal buildings around the world with new, more energy-efficient designs and technology. Even private buildings — such as the Sears Tower — and more than 20 colleges and universities in the United States are undergoing energy-saving retrofits by partnering with the Clinton Climate Initiative. Even further, over 25 companies have agreed to offer lower prices for their energy-efficient materials, systems, and other products — making them more widely accessible to more than 1,100 cities around the world through the Clinton Climate Initiative.

The importance of youth

Youth play an important and unique role. They can add a real sense of urgency to climate talks by stressing that the future — which may seem abstract to policymakers, industry, and the public — is very real for them. When scientists say, "This is what the world is going to be like in the year 2050," this young generation will live through those major changes. The threats are very real, and so is the need to immediately implement solutions.

The most useful element that youth bring to the table is optimism and high energy. They also invoke a sense of moral obligation in their elders. The involvement is genuine and fresh, and the lack of years of experience means that most youth see climate change with a sense of simplicity. When youth speak about global warming, they do so without employing numbing jargon or invoking complex political issues. What matters to them is their future, and they say so purely and simply.

Groups that speak up

In just the past handful of years, coalitions and networks of youth organizations working on climate change issues have formed. They're sharing resources, organizing networking meetings between youth climate groups, and bringing youth representatives to UN climate change conferences. Here are just a few of those groups:

- **The Australian Youth Climate Coalition (`www.youthclimatecoalition.org`):** This group is gaining strength and members quickly. It offers an educational program called Switched on Schools that helps high school students become active in their schools and communities, and it ran a campaign called Adopt a Politician, which encouraged youth to lobby their representative to act on climate issues leading up to the 2007 federal election.
- **The Energy Action Coalition (`www.energyactioncoalition.org`):** This group, of which Zoë is a member, brings together almost 50 youth-run organizations across the U.S. and Canada. Not restricted to climate issues, the coalition has an array of foci, ranging from environmental justice and politics to community and education. It runs the Campus Climate Challenge program, which implements clean energy policies and emission reduction strategies at colleges and universities while raising awareness among students. This program already has more than 570 campus groups involved.

  The Coalition was the brainchild of Billy Parish, who dropped out of Yale to do more important things, like save the planet. In 2007, the program was featured in *Vanity Fair*'s green issue, which earned Zoë and her team of colleagues a spot in the magazine's photo spread — on the page before Robert Redford!
- **Solar Generation (`www.solargeneration.org`):** The youth arm of Greenpeace operates in over 20 countries around the world. They're working on more than 120 solar panel installation projects in Switzerland, convincing Australian and American universities to begin installing on-campus solar panels, organizing skill-share energy and climate conferences, and participating in peaceful climate protests.

Getting Involved

Because many non-governmental organizations rely on volunteers, they're always looking for people to join the team. If you're looking to get involved in climate change issues, participating in an NGO is one of the best things you can do. You can make a difference, especially as part of a larger group. As a member of an organization, you can rely on a support network of people who have experience addressing climate change issues; they can help you find where you can best contribute.

Seeking out groups

As we show in the section "Understanding What Non-Governmental Organizations Do," earlier in this chapter, a wide array of NGOs exist. One's bound to match your interests. Here are some suggestions about how to find a group that's right for you:

- **Attend a conference or event.** These hotspots for meeting representatives from NGOs can help you find out about each group's projects and how you can get involved.
- **Contact your area's environment department.** Often, your provincial, state, or federal government's department of the environment has lists of organizations working on climate change. (We share some government Web sites that can point you in the right direction in Chapter 22.)
- **Show up at Green Drinks.** Starting off as a good idea by a small group of friends in 1989 in the U.K., the event Green Drinks now occurs monthly in over 30 countries around the world, and in more than 350 towns and cities. People from academia, business, government, and non-governmental organizations who are working or interested in the environmental field meet up. It's a great place to meet the inner circle of climate change–savvy people in your area and get connected with groups acting on the problem. Green Drinks is widespread across the U.S., the U.K., Australia, and Canada. Check out `www.greendrinks.org` for more.
- **Surf the Web.** The sidebar "Joining the climate change team(s)," in this chapter, lists the Web site addresses of some of the major NGOs. Many of these larger organizations have local chapters or can at least direct you to groups in your area.

Helping out

Organizations often have multiple campaigns going on at the same time and always appreciate help at any level. You can get involved in many ways, depending on your skills and interest.

Most organizations have Get Involved or What You Can Do sections of their Web sites that link directly to current projects for which they need help. Here are some examples of volunteer positions:

- **Fundraisers:** Raising money for organizations working on climate change ensures that they have funding to keep fighting climate change.
- **Organizers:** If you have strong organizational skills, many groups will be more than happy to have you help organize (or even take on organizing) public awareness events or conferences on climate change in your community.
- **Public speakers:** Speaking at events, big or small, is a help to organizations. Often, organizations offer training, so that you know just what to say. (In Chapter 19, we look at how you can get involved with the Climate Project, a group that trains people to offer the same presentation that Al Gore gives in the documentary *An Inconvenient Truth.*)
- **Writers:** Organizations are always on the lookout for people to write letters to government, business, and industry representatives. Writers are also a big help in creating press releases and articles to get issues into the public eye through the media.

Joining the climate change team(s)

Many great groups are working on climate change issues. Some focus only on how to make religious buildings more energy efficient, some concentrate on greening schools. Some are active on a local scale in their own community or region, and others are taking it global. Here's a tiny snapshot of major NGOs in Australia, Canada, the U.K., and the U.S. that focus on climate change:

- **American Solar Energy Society (`www.ases.org`):** Active since 1954, this organization promotes and implements the use of solar energy, energy efficiency, and other renewable energy technologies. You can join a local chapter and receive guidance on installing solar energy in your home.
- **Clean Air — Cool Planet (`www.cleanair-coolplanet.org`):** This American organization works with companies, campuses, communities, and science centers to cut greenhouse gas emissions and engage people in implementing their own climate solutions. This organization can work with you to join local groups, find the main problems that your area needs to deal with, build a plan, and then implement it.
- **David Suzuki Foundation (`www.davidsuzuki.org`):** A large Canadian organization led by icon environmentalist David Suzuki, this group offers a Nature Challenge to members to reduce their carbon footprint and work together to reduce our emissions.
- **Friends of the Earth (`www.foei.org/`):** This organization has over 2 million members and 5,000 local chapters in 69 countries. Members volunteer locally with chapters to reduce the

(continued)

(continued)

greenhouse gas emissions of their communities, and many members also get involved by responding to requests from the Friends of the Earth headquarters — such as writing to their government representatives on an urgent climate-related issue.

- **Green for All (www.greenforall.org):** This organization's main campaign is to create an economy for green collar jobs in the United States, enabling people to make a living making the Earth more livable. Green for All provides volunteers with resources to campaign in their own communities for more environmentally friendly job opportunities.
- **Greenpeace (www.greenpeace.org/~climate):** Greenpeace is a major international organization with many climate change awareness programs: You can sign up for a seven-step climate change challenge, or you can become a Cyber Activist — receiving online alerts of when and how to take action. Greenpeace accepts all kinds of volunteers; they work with the skills you have and fit you where you can be most effective.
- **Sierra Club (www.sierraclub.org):** What began in 1892 as a club dedicated to bringing people into the great outdoors to experience the wonder of nature has grown into one of the strongest environmental groups and advocates for battling climate change. You can join in Canada and the U.S. and belong to a local chapter or youth branch, go on outdoor outings, and volunteer for campaigns. You can find Sierra Club Canada at www.sierraclub.ca.
- **Union of Concerned Scientists (www.ucsusa.org):** This organization, founded and staffed by scientists, promotes a high level of public understanding and practical science-based solutions towards a better world. Anyone can join and contribute to the discussions and research that shape the group's reports and policy recommendations. The UCS has been active for over 35 years throughout the United States, addressing critical issues, such as climate change.
- **World Wildlife Fund (www.worldwildlife.org/climate):** WWF is active all over the world. When you become a member, WWF continually updates you on WWF climate campaigns and how you can get involved in international climate conferences, gives tips for making green changes at home, and offers partnerships to green your business. If you live near a WWF office, you can volunteer on-site, too.

Chapter 16

Lights, Camera, Action: The Media and Global Warming

In This Chapter

- Reviewing the treatment of global warming in the news
- Looking at how Hollywood and its stars are responding to climate change
- Checking out climate change blogs
- Curling up with a few global warming books

Not too long ago, global warming was the exclusive domain of climatologists and environmentalists. Now that its effects are undeniable, it's on everyone's lips and getting serious exposure in popular culture. This is good news, as far as we're concerned — the more people know about climate change, the greater the momentum for change.

But is any publicity good publicity? In this chapter, we take a close look at how the news and entertainment industries cover global warming and show how the science that we cover in Part I sometimes gets lost in the push for a good story.

Growing News Coverage

Once relegated to articles in science journals and the back pages of the newspaper, global warming now merits cover stories in mainstream magazines, front-page stories in the papers, and feature reports on news broadcasts.

Although the number of climate change stories is increasing worldwide, the rate of growth is different from nation to nation. An Oxford University study revealed that the British newspapers offer far more stories on global warming than their U.S. counterparts. Check out Figure 16-1 to compare the increase in newspaper coverage on climate change from 2003 to 2007 in the U.S. and the U.K.

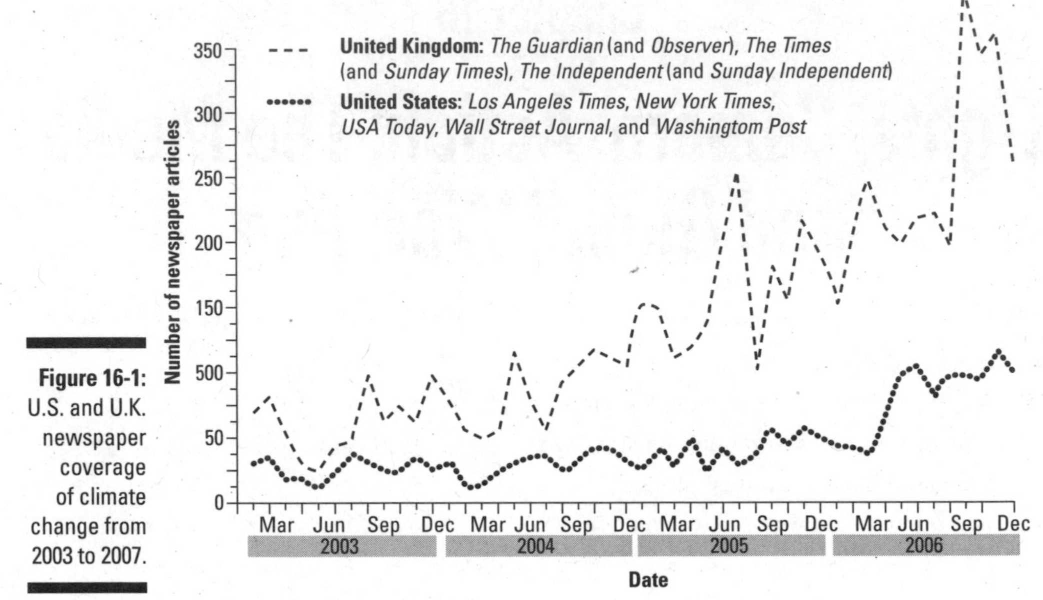

Figure 16-1: U.S. and U.K. newspaper coverage of climate change from 2003 to 2007.

Max Boykoff. "Flogging a Dead Norm?" Area *39:4.*

Without this coverage, public awareness of global warming wouldn't be where it is today. But, as we cover in the following sections, you need to read those feature stories and watch those reports carefully, no matter where they're published or broadcast, to ensure you don't get misinformed.

Bias and balance: Distorting the story

Because journalists are supposed to give both sides of a story, sometimes they actually create a bias in their reporting. In 2004, academics Jules and Maxwell Boykoff published a study of more than 600 randomly selected articles on climate change that appeared in *The Wall Street Journal, The New York Times, Los Angeles Times,* and *The Washington Post.* Just over half the articles gave roughly the same amount of attention to competing views: the idea that humans contribute to global warming and the opposite claim that climate change is exclusively the result of natural fluctuations. The academics found that television coverage was split more evenly between the two perspectives, with 70 percent of television news coverage giving both sides equal airtime.

Although this "balanced" reporting might seem fair, the likelihood that humans are contributing to climate change is 95 percent certain, according to the Intergovernmental Panel on Climate Change (IPCC). Giving both sides equal coverage creates the inaccurate perception that it's an equally weighted debate. Similarly, the Boykoffs' study found that 78 percent of the articles were balanced in reporting competing views about acting on the threat — either act now, or wait and see. Again, though, the science overwhelmingly advises people to act now.

Consider the source: Being an informed media consumer

You sometimes have to really work to figure out when something you read or see in the media is 100-percent true. But by being a careful media consumer, you can tell the difference between credible information and inaccurate reporting. Watch out for these key issues when reading articles or watching features about global warming:

- **Questionable experts:** Television news reports need talking heads, and newspaper articles require people who can act as sources — and plenty of people are willing to offer themselves as experts, with or without the knowledge required. A handful of "experts" always support the climate denier side of the argument. These pundits like media coverage and appear to have good credentials, but they may not actually do research on climate, they may receive funding from fossil fuel companies, or they may distort the work of others. Media-friendly experts who stress immediate action on global warming can be problematic, too. Although they feel passionately about the issue, these individuals may communicate the wrong information by oversimplifying complex science or exaggerating global warming's effects.

  You can tell whether or not these people are experts by checking to see whether the news story references any scientific research that the expert has actually had published in a peer-reviewed journal.

- **Old research or evidence:** If the report doesn't mention the date of the evidence it's citing, that's a red flag — the evidence could be an old theory that has since been refuted. The documentary *The Great Global Warming Swindle,* for example, focused on solar cycles as a cause of global warming, using evidence from the 1980s that has since been largely disproved. (See the following section for more about this documentary.)

- **Source of information:** If a report references the IPCC, you can trust that information — as long as the writer or producer hasn't misunderstood the words of the IPCC. In his novel *State of Fear* (HarperCollins), Michael Crichton quoted IPCC out of context a fair bit! (See the section "Bestselling Books: Reading between the Lines," later in this chapter, for more about this book.) A scientist who has published and peer-reviewed articles offers another good source of reliable information.

Blogs and Wikipedia, on the other hand, are generally the least credible sources of information about climate change. Anyone can create or contribute to these Web sites. (Nevertheless, we look at a few of the most popular blogs in the section "Worldwide Warming: Climate Change Blogs," later in this chapter.)

For reliable online information, try government, non-governmental organization, and university Web sites. (We recommend ten of our favorite resources in Chapter 22.)

- **Oversimplification:** News reporters often simplify information to a point where it's actually incorrect or over-exaggerated. For example, if a scientist says sea levels could rise by 13 to 16 feet (4 to 5 meters) if the Western Antarctic Ice Sheet collapses, a reporter might say the scientist has *predicted* this will happen. (Predicting and projecting are two different things. A *prediction* is an assertion that something *will* happen, but a *projection* is an extrapolation of future events if current trends hold true.) Change the trend lines and the projection will change.

  Complex scientific issues don't fare well in the land of the sound byte. If the article you're reading or the report you're watching offers extreme outcomes, check for further background information to ensure you get the whole story.

- **Surveys:** Surveys are a great way to gauge public opinion — if they're accurate. When reading about a poll, check how many people were surveyed. A small number of people (say, 50) doesn't necessarily mean that the information is wrong, but the poll or survey results probably don't reflect the greater population.

Science on the Red Carpet

Climate change is so hot that Hollywood studios are taking notice. And their stars are acting, as well — not just on-screen, but in their personal lives. But what's the real story? How much truth is there in *The Day After Tomorrow*? How green — really — are those jet-setting stars? And can George Clooney make a difference by tooling around Tinseltown in an electric car?

Movies: Facts and (science) fiction

Although global warming is constantly gaining public attention, it hasn't served as the subject (or even the backdrop) for many films. Because global warming poses a complex problem that requires complex solutions, it doesn't make for the stuff of Hollywood fodder. For the most part, when climate change has turned up on the silver screen, its portrayal has been sensationalistic, setting the scene for science-fiction dystopias. Happily, some documentaries set the story straight, and some of the best work that's being done on the screen is explaining the problem to kids, who will be on the front lines of climate change impact while they grow up.

Silver screen stories

Perhaps the most prominent fictional film that addresses global warming, *The Day After Tomorrow* (2004), is a well-intentioned movie that's scientifically credible . . . for the first 15 minutes. The movie opens promisingly — the film's hero, a climate scientist played by Dennis Quaid, briefs a world conference on his theory: When Arctic ice melts, the fresh water released will interfere with the oceans' thermohaline currents, slowing them down and leading to climatic disasters. (Chapter 7 discusses the actual computer modeling of this possibility.) The fictional U.S. vice president reacts realistically, demanding to know who would pay the costs of Kyoto (see Chapter 11).

But after this believable beginning, the plot becomes completely driven by special effects, and the real science is lost in the (literal) deluge. We can't deny the coolness of watching freak tornadoes obliterate Los Angeles, snow pummel New Delhi, and killer hail devastate Tokyo — but by the time a deep freeze destroys the entire northern half of the planet, we're no longer noting scientific accuracy. Although special effects dominate the movie, the producers acknowledge the liberties taken and include useful interviews with scientists in the DVD's special features.

Science fiction films dominate the list of fictional movies that deal with global warming. In terms of quality, they're a mixed bag:

- ***A.I.: Artificial Intelligence* (2001):** This post-disaster film takes place after all the polar ice caps have melted and most coastal cities are entirely flooded, pushing civilization inland. Although humanity might never see a robot with feelings, which is the subject of Steven Spielberg's film, flooding of low-lying cities is a realistic possibility. If people keep producing emissions at the current rate, coastal waters could rise 8 to 24 inches (20 to 60 centimeters) by the year 2100.
- ***The Arrival* (1996):** This little-seen movie is heavy on suspense but light on science. It turns out that global warming is the work of aliens, who have come to Earth and are working secretly to transform the planet to suit their own needs.
- ***Waterworld* (1995):** Kevin Costner starred in the ill-fated *Waterworld* as a mariner sailing the Earth after global warming triggered flooding, which has covered almost all land. The only ground still above water, called Dryland, turns out to be the summit of Mount Everest. Sadly for

> the movie's credibility, the world doesn't have enough water to flood the whole thing. The highest estimates of how far sea levels could rise from global warming max out at about 650 feet (200 meters). That's a lot, but the *Waterworld* scenario would need something like 30 times that depth.

One rare exception to the special-effects-driven Hollywood take on climate change is Rob Reiner's *The American President* (1995). This romantic comedy uses the climate issue as an effective backdrop without hitting a single false note on the science. The film depicts the unfolding love story between the president (a dishy widower played by Michael Douglas) and an earnest lobbyist (Annette Bening) for a big, mythical environmental group. The lobbyist isn't satisfied with the president's commitment to cutting energy use by 10 percent, even back in the mid-1990s. With no special effects required, the script stays relatively error-free and helpful — for example, when the president gives his lover his phone number, it's the actual White House line.

Documentaries

The most influential full-length movie on global warming to date is the documentary *An Inconvenient Truth* (2006), featuring former U.S. Vice President Al Gore. Building momentum through word-of-mouth and low-scale advertising, it was one of the year's most successful films — an unheard-of feat for a documentary. The buzz grew even louder when the film walked away with the Academy Award for best feature-length documentary.

An Inconvenient Truth is based on Gore's lecture presentations that highlight big, catastrophic events that global warming may cause. Called a true horror film by some for the impact it has on viewers, it makes the scale of the problem clear to audiences — that the situation is way beyond such easy fixes as switching light bulbs. However, it does outline what individuals can do, and it definitely motivated many viewers. The movie essentially changed global thinking, moving climate change to the forefront of many people's minds. To follow it up, thousands of "mini-Gores" are being trained to give the same presentation across the United States. Gore himself, along with the Intergovernmental Panel on Climate Change (IPCC), was awarded the Nobel Peace Prize in 2007 for his climate change work.

Another strong documentary is *Too Hot Not to Handle* (2006). The program covers the effects of global warming on the United States. It was originally aired on the U.S. television channel HBO and, like *An Inconvenient Truth,* is now available on DVD.

The Great Global Warming Swindle, screened in 2007 in the U.K., presents a far more contentious view of global warming, arguing that civilization's impact on climate change has been overstated. Martin Durkin, the documentarian, claims that natural trends, such as solar cycles, cause global warming. (We talk about solar cycles' contribution to global warming in

Chapter 3.) Many critics note, however, that scientific research has confirmed the impact of human activity on global warming with greater certainty than the film acknowledges. In fact, recent research determined that the sun's activity has actually been decreasing since 1985, making it all but impossible that solar cycles are causing the planet's warming. Chapter 21 dispels some of the myths that this film presents.

For the kids (and adults, too)

You might think that global warming is unlikely subject matter for children's entertainment, but a couple of recent kids' movies have tackled climate change:

- ***Ice Age 2: The Meltdown* (2006):** This movie makes good-humored references to the natural Ice Age and warming cycles while touching on such current threats as flooding. Roger Ebert noted, "If kids have been indifferent to global warming up until now, this *Ice Age* sequel will change that forever."
- ***Happy Feet* (2006):** This environmentally minded movie features penguins who are dealing with the consequences of civilization's encroachment on the natural world. Although the movie doesn't expressly address global warming, it does get kids thinking about the impact of society's actions on the greater environment.

Following the stars

"Well, if Leonardo DiCaprio is doing it . . ." Okay, maybe you need more motivation than that to get involved, but the power of the stars is undeniable. Many celebrities are using their high profiles to advance green living and advocate action on climate change. DiCaprio and others are speaking out, getting involved with climate change organizations, and practicing what they preach by making smart choices about cars, travel, and home retrofits. Not everyone can spend hundreds of thousands of dollars on high-tech goodies, but everyone can do his or her part, and these stars deserve credit for making these choices and encouraging others to do the same.

The organizations that stars work with get a lot of press coverage and free publicity. Concerns exist about turning an urgent issue such as climate change into a pet cause — after all, a celebrity could be ill-informed, and the media would print his or her words anyway — but the benefits far outweigh any concerns. Most of the time, the stars know what they're talking about, and they're genuinely concerned about the issue. In this kind of situation, the ideals associated with celebrities (as they are so often in the media limelight) can influence people — and leaders — for the better.

Leading lights

Here are just a few of the stars putting their celebrity to work:

- **George Clooney:** Twice named the Sexiest Man Alive by *People* magazine, he just got sexier. He drives the ultra-mini, one-seat-wide Commuter Car Tango, and the Tesla Roadster, a top-of-the-line electric beauty.
- **Leonardo DiCaprio:** Climate change is a passion for this actor, who co-wrote and produced an environmental documentary, *The 11th Hour.* He also created the Leonardo DiCaprio Foundation in 1998 to support a variety of environmental causes. DiCaprio sits on the boards of the National Resources Defense Council and Global Green USA. In his personal life, he has installed solar panels on the roof of his house.
- **Jack Johnson:** This professional surfer turned musician hosts an annual music festival close to his home in Hawaii with all the proceeds going towards the Kokua Foundation, which promotes environmental education.
- **Willie Nelson:** Following up his foray into marketing "BioWillie," a blend of 20 percent biodiesel and 80 percent regular diesel, the veteran country singer has thrown his support behind the not-for-profit Sustainable Biodiesel Alliance.
- **Brad Pitt:** Another former Sexiest Man Alive, Pitt is working with the Home Depot Foundation to build the first totally self-powering house as part of rebuilding New Orleans after Hurricane Katrina.
- **Julia Roberts:** The *Pretty Woman* star uses environmentally friendly products (including her kids' diapers), has solar panels on the roof of her home, and uses bio-fuel in tractors and equipment at the family ranch. She's a spokesperson for Earth Biofuels, Inc., and chairs its Advisory Board. She's also involved in a community project to protect 100,000 acres of wildlife habitat in New Mexico's Valle Vidal.
- **Arnold Schwarzenegger:** Famous for his role as *The Terminator,* Schwarzenegger quit his acting job to run successfully for governor of California. He's been a tireless advocate for action on climate change. (Check out Chapter 10 for some of his successes.)

Perhaps the most inspiring star is Robert Redford. Redford created Sundance Village, a haven for environmental art and conservation, on land he bought in the late 1960s in the heart of the Utah mountains. This initiative gave birth to the annual Sundance Film Festival, where *An Inconvenient Truth* had its world premiere, and the Sundance Channel, which broadcasts a three-hour, prime-time programming block titled *The Green,* devoted entirely to the environment. Actively fighting for conservation for decades, Redford has been described as a steadfast, well-informed voice on environmental issues.

He founded the not-for-profit Institute for Resources Management to try to bridge the gap between environmentalists and industry, hosted a Sundance conference on global warming way back in 1989, and started the conservation initiative North Fork Preservation Alliance to promote land stewardship.

Carbon offsetting and green travel with the jet set

One area where celebrities don't often set a great example, however, is in their jet-setting ways. Some environmentalists give performers a hard time for the amount they travel by plane (often private), and for the extensive energy and fuel needed for major press junkets, music tours, and personal trips. Fortunately, many celebrities are beginning to carbon offset their air and ground travels — and the list is long, ranging from Tom Cruise to Nicole Kidman to Steven Spielberg. (Check out Chapter 17 for information about carbon offsetting.)

Although performing artists still tour, many of them try to make up for it along the way by fuelling their buses with biodiesel blends, carbon offsetting, and providing information on environmental issues to their fans. The leaders in this movement include Bonnie Raitt, Willie Nelson, Jack Johnson, Alanis Morissette, the Barenaked Ladies, and the Dave Matthews Band. In fact, the Barenaked Ladies have their own carbon-offset organization, dubbed Barenaked Planet.

Worldwide Warming: Climate Change Blogs

Blogs are booming. A *blog* (short for Web log) is an unofficial article or opinion piece posted by an individual on the World Wide Web. Anyone surfing online can access the blog and respond by posting a comment. Most blogs aim to spur back-and-forth discussion among readers and with the author.

Here are a few of the most-read, climate-centered blog sites:

- **AccuWeather** (`http://global-warming.accuweather.com`): This blog offers global warming news, science, myths, and articles. It's run by a senior meteorologist — at the time we wrote this, it's Brett Anderson, a meteorologist with more than 18 years experience.
- **Celsias** (`www.celsias.com`): This blog offers practical solutions and climate change projects you can take on — such as responsible tourism, creating off-grid engines, and green home construction. The site's slogan is certainly inviting: "Climate change is not a spectator sport."

- **Climate Ark** (www.climateark.org): This blog features a timely news feed and focused discussion topics. It also includes links to many other blogs and climate resources.
- **Climate Feedback** (http://blogs.nature.com/climatefeedback): Sponsored by *Nature* magazine, this blog aims to raise questions and debate about climate change issues.
- **Climate Progress** (www.climateprogress.org): This blog focuses on U.S. politics, climate science, and — as its name conveys — progressive solutions.
- **Desmog Blog** (http://desmogblog.com): This blog hits on a wide range of climate change issues, with the goal of removing the "smog" of media hype.
- **It's Getting Hot In Here — Dispatches from the Youth Climate Movement** (www.itsgettinghotinhere.org): The posts on this relatively new blog come from young people, from students to activists to professionals, posting on climate issues (mostly in the U.S. and Canada). Zoë posts on this blog, which now hosts more than 60 regular contributors.
- **Real Climate** (www.realclimate.org): Although some of the language on this blog is science-based, the climate scientists who post here are writing for the public. They aim to stick to scientific discussion and stay out of politics and economic issues.
- **World Climate Report — The Web's Longest-Running Climate Change Blog** (www.worldclimatereport.com): This blog aims to give both sides of any issue being discussed. Despite this, *Nature* magazine has referred to it as the "mainstream skeptic" point of view. Translation: Read with caution!

Although not strictly climate-change-focused, TreeHugger (http://treehugger.com) and Gristmill (http://gristmill.grist.org) thoroughly cover global warming issues. Check them out for a slew of well-written articles.

Blogs can often be the source of the *least* credible information. You have to judge the legitimacy of the writing, which depends entirely on the author, his or her background, and his or her research. Remember, anyone with access to a computer can post a blog or comment on one.

When you're in doubt about a blog's credibility, click on the About Us link that most sites feature. The bio page that appears often reveals just how legitimate the blog's information is. If the individual or group has a clear bias, you can probably guess the influence that bias has on the information you're reading.

Bestselling Books: Reading between the Lines

The number of books on climate change has exploded in recent years. From science to fiction, fear to solutions, and children's books to adult titles, you have plenty to pick from.

True stories

A number of extremely well-written, non-fiction books on climate change have appeared in recent years. Some are long on the problem and short on offering solutions, which can be . . . well, a little depressing. Among the better-known titles are

- ***Field Notes from a Catastrophe,* by Elizabeth Kolbert (Bloomsbury):** Based on Kolbert's articles for *The New Yorker,* this is a first-person journalistic look at the science behind and the impacts of climate change. Kolbert offers a well-researched and clearly explained account of the urgency of climate change while linking this to real places and stories from around the world.
- ***Heat,* by George Monbiot (South End Press):** Monbiot, one of the U.K.'s most respected journalists, offers a truly radical approach to avoiding atmospheric tipping points. His writing reflects the immediacy of the climate crisis by demanding changes such as rationing energy use. Monbiot doesn't believe that the range of policy options we describe in Part IV can get greenhouse gas levels low enough fast enough. He could well be right. (See Chapter 20 for more about Monbiot.)
- ***An Inconvenient Truth,* by Al Gore (Rodale Books):** Tied to the film of the same name (which we discuss in the section "Science on the Red Carpet," earlier in this chapter), this current affairs book is the most accessible title on climate change currently available. Gore conveys his message with minimal text, using easy-to-read graphs to show the science behind climate change.
- ***The Weather Makers,* by Tim Flannery (Atlantic Monthly Press):** Beautifully written, this book covers environmental science and issues in detail and depth. Though the content is a little overwhelming, Flannery offers solutions and a vivid writing style that draws you in. This book is a little frightening, but offers solid and entirely correct scientific information. (We talk more about Tim Flannery in Chapter 20.)

- ***The Winds of Change,* by Eugene Linden (Simon & Schuster):** Based in science and history, this book covers such overarching topics as the Gulf Stream, El Niño, weather patterns, and temperature. It gives you a very good overview of the chronology of climate change science, politics, and debate. It provides little in the way of solutions; however, it is well written, interlaced with personal, historical, and political anecdotes.

Fiction and fairytales

Global warming makes a good story, tempting writers to use it as a basis for fiction. Like with movies (see the section, "Science on the Red Carpet," earlier in this chapter), however, the underlying science isn't always presented properly. First, here are a couple of solid efforts:

- ***Floodland,* by Marcus Sedgwick (Yearling):** This imaginative book geared towards young teens made Zoë a little nervous. It's about a young girl named Zoe who's stranded in a rowboat in a flooded world.
- ***A Scientific Romance,* by Ronald Wright (Picador):** Inspired in part by H. G. Wells's classic *The Time Machine,* Wright's hero travels 500 years into the future, to a Britain transformed into a depopulated tropical jungle thanks to global warming. The author's beautiful prose and deft description make the situation seem all too plausible.

One novel we don't recommend for satisfying your climate change curiosity is *State of Fear* (HarperCollins), a 2004 thriller by Michael Crichton. The book depicts global warming as a conspiracy concocted by conniving environmentalists. The story would be amusing if some people didn't take the book seriously. The chair of a U.S. Senate committee invited Crichton to testify on matters surrounding research being used for public policy, and Crichton visited the White House to chat with President George W. Bush, who had read the book. The American Association of Petroleum Geologists (which gets funding from the petroleum industry) even gave him their annual Journalism Award.

Crichton says that *State of Fear* is the product of three years of research, offering detailed footnotes and an appendix to support his claims. Unfortunately, he misinterprets and misrepresents much of the science. Crichton even attacks real scientists through his novel, with one character claiming that a prediction made by eminent U.S. scientist Dr. James Hansen 1988 about rising temperatures was off by 300 percent. Hansen himself has refuted that claim, showing that his projection was in fact remarkably accurate — an "inconvenient truth" for Crichton. (Check out Chapter 20 for more about Dr. James Hansen.)

Chapter 17

Taking the High Road

In This Chapter

- Getting around without a car
- Driving your car with fewer emissions
- Flying green through the blue skies

The area in your own life in which you can best limit greenhouse gas production is transportation. Any means of transportation that relies on fossil fuels adds greenhouse gases to the atmosphere — and currently, that's just about every means of transportation. No wonder transportation accounts for about a quarter of all human-produced greenhouse gases.

People can and should push governments to change transportation policies to move citizens away from car dependence. Investments in mass transit within cities and efficient rail systems between cities can help many people make a choice to leave the car at home — and so can pricing policies that create incentives for low-carbon travel (a carbon tax sends a message to all of society that wasting fossil fuels wastes money). (Check out Chapter 10 for more about what governments can do.)

But the decision to walk to the corner store instead of driving, to buy a low-emissions car rather than a big SUV, or to take the train rather than a plane is yours and yours alone. In this chapter, we look at the travel options you have, whether you're making a quick jaunt or taking a long haul, and we highlight the greenest choices.

Opting Out of Automobiles

Modern industrialized society's cars are such a part of life that some people living in these societies may have a hard time imagining making even the smallest trip, such as a jaunt to the corner store, without them. But those little trips add up and contribute to greenhouse gas emissions.

For every mile that you drive, a conventional car powered by an internal combustion engine releases one pound of carbon dioxide into the atmosphere from the tailpipe. So, how do you get from Point A to Point B without adding to global warming? You have a lot of options, which we explore in the following sections.

Choosing where you live

Living close to mass transit or within walking distance of where you work, shop, and send your kids to school can help reduce your transport-related emissions. According to the American Public Transportation Association, Americans who live closest to public transportation drive 4,400 miles (7,081 km) less a year than those who don't have easy access to buses or trains. That's a lot less exhaust.

Stepping in the right direction

Walking is the easiest choice for short distances. Walking can take just about as much time as starting up the car, waiting at traffic lights, and searching for a parking space — especially if you're going only a few blocks. And your brisk jaunt will be carbon-free!

Depending on where you live, work, or go to school, you can probably use your feet for short trips every day, instead of driving. Try walking when you

- Run errands between places downtown (instead of driving from shop to shop).
- Go to lunch or dinner.
- Shop for food (bring a backpack or personal shopping cart, as well as your reusable shopping bags).
- Go to work if you live within about a half-hour walk.
- Go to the gym (bonus — you can skip your warm-up!).

Putting the pedal to the metal (of your bike, that is)

The bicycle is the most efficient mode of human transportation ever invented, consuming the least energy per mile traveled. Bikes are about 5 times more efficient than walking and 50 times more efficient than cars.

Bicycles take up far less room on the roads, reduce traffic congestion and smog, and improve their riders' health. Biking does have it's challenges — it may not be an all-season option in many places, and some trips are too long for a bike ride. Despite these limitations, riding a bike when possible, rather than taking a car, is the environmentally friendly choice. If you want a little more speed, opt for a moped or scooter (see Table 17-1 for information about scooter fuel efficiency).

Taking mass transit

Mass transit was defined in one of Elizabeth's university classes as "the conveyance of persons in bulk." Like the name suggests, it's most efficient when it carries a large number of passengers.

Tips for top biking

If you're interested in giving carbon-dioxide-free transport a spin but don't know a unicycle from a tandem bike, these hints can help set you on the right path:

- **Choose a bike that's right for you.** Select the kind of bike you want based on your needs, your fitness level, the kind of riding you plan to do, and even the local weather and landscape. You have many different styles and sizes to choose from. Visit a bike store near you and talk to an expert about your bike needs.

  If you have to navigate the heavily used streets of some cities' downtown areas, you might also want to get tires that are lined with Kevlar to resist punctures from broken glass.

- **Get the proper accessories.** After you have the bike, you need a helmet, a lock, a bell or horn, and lights so that cars can see you after dark. (In many places, the law requires you to have these items.)

- **Chart out bike-friendly routes.** If your city hasn't yet adopted separate bike lanes or bike trails through built-up areas, you can get around more safely by following less-busy streets that run parallel to the major routes.

- **Take your bike on the bus.** Progressive policies in some cities include allowing bike riders to take their bikes on the bus — usually on a rack along the front of the bus. By taking a bike on public transportation, a bike rider can use that bike on the parts of the route that buses don't service. The cities that have bike racks available on the bus, or attached to the front of the bus, encourage this *intermodal* (involving more than one mode of transportation in a single trip) approach to transportation.

- **Be a concerned cyclist.** If your city or town doesn't offer dedicated lanes or paths for bicyclists, contact your local government and encourage officials to consider adding them.

Go car free . . . for a day

The first car-free days were held in Switzerland in 1974 in response to the 1970s Oil Shock. Major car-free events in Reykjavik, Iceland, and La Rochelle, France, in the late 1990s led up to the official creation of Car Free Day in 1998. To coincide with European Mobility Week, Car Free Day was held on September 22, and still is in many cities around the world. This grass-roots movement has spread around the world. In some cities, the streets are actually closed to all traffic for the day.

Rome was one of the 150 cities and towns across Italy that went car free for a day in 2000, bringing life and a thorough sense of enjoyment to the people-busy streets. In other cities, certain neighborhoods close to traffic in a more symbolic celebration. Athens, Paris, Madrid, and Hamburg see a significant drop in cars on Car Free Day. Dublin closes three streets. Toronto celebrates largely in the Kensington Market area by closing the area to cars and bringing vendors out into the street, creating a social market atmosphere. The city of Bogota in Columbia has the largest Car Free Day, during which the municipal government completely restricts the use of vehicles that run on fossil fuels. Over 500 cities participate in annual car-free days held on Earth Day, April 22.

If you want to reduce greenhouse gas emissions and can't walk or bike to your destination, mass transit is the best way to travel. You take up a seat on a vehicle that's running anyway — so your trip costs almost nothing in terms of additional fuel consumption, emissions, and pollution.

Riding the rails

Trains are the best option for long-distance travel: they're safer than cars and more comfortable and more fuel efficient than planes (cheaper, too, in most cases). Traveling by train can be much more fuel efficient than driving by yourself in your car, and the efficiency rises when more people get on board.

Added bonuses of taking the train, especially on longer trips, are that you can sleep in comfort (even in your own room), get up and move around easily, and use the extra time to get things done without a lot of distractions. (In fact, we wrote over half of this book on a train!)

When calculating your carbon footprint, train travel may seem like a high-carbon option. A higher carbon footprint for riding trains reflects low passenger usage in some countries because nearly empty trains can't be very efficient. Ridership is much lower in Canada and the United States, for example, than in the European Union or Japan. Take the train to help improve your country's results! (Flip to Chapter 18 for information on where and how to calculate your carbon footprint.)

Going the distance on the bus

Buses are a very environmentally friendly way to travel. The good old Greyhound (the North American bus service) could change its slogan from "Go by bus and leave the driving to us!" to "Go by bus and reduce your ecological footprint!" (Not as good, we know — there's no rhyme, but plenty of reason.)

Traveling by bus emits five times fewer greenhouse gas emissions than traveling by plane. It's among the most environmentally friendly ways you can travel. Bus routes often reach the most remote of communities, whereas trains tend to follow a main city route, stopping at only whatever lies in between.

Greyhound is the main coach bus service provider in the U.S., Australia, and Canada. Greyhound Australia runs its buses on the lowest-emission diesel fuel available and has spent over $10.5 million U.S. in the past year on upgrading to a fuel-efficient fleet of buses.

National Express is the go-to bus provider in the U.K. This company offers passengers a carbon calculator and is a proud member of The Climate Group's We Can Solve It campaign, a global initiative to bring organizations, businesses, and individuals together to fight climate change.

Taking public transit

Depending on where you live, you might have a decent public transportation system — and if you do, hop on board! Although taking the city transit has been stigmatized as crowded and unpleasant, we can't think of anything more unpleasant than helping heap carbon dioxide into the atmosphere.

On track with global train travel

The popularity of train travel varies around the world. In Europe and Japan, train technology is much more advanced than in the rest of the world — their trains are fast, convenient, popular, and affordable. Rail Australia is also well developed and functions in almost every region of the country.

In North America, trains are far less popular — and therefore far less efficient — but their popularity is growing, thanks to high gas prices and airport security concerns. Amtrak, the national U.S. passenger train company, offers the only train service in North America that comes close to the high-speed trains of Europe, the Acela. Travelers can get from New York to Washington, D.C., in about two and a half hours, making it a popular alternative to air travel. In Canada, where ViaRail operates the national passenger rail system, ridership has been growing steadily.

The two main forms of public transportation in most cities are the bus and subway. Most buses run on diesel fuel, but many have already been upgraded or replaced with biodiesel, fuel cells, or electric motors, so they're even more greenhouse gas–friendly. The subway (called the Metro, Underground, Tube, and so on, depending on where you are) runs on electricity, so its emissions depend on the source of the electricity. Efficiency-wise, it's at its best when fully loaded, like any form of mass transit. Adding to the subway's fuel efficiency is the lower friction of train wheels on metal tracks, compared to vehicles that have rubber tires that run on road surfaces. The less friction the wheels encounter, the less energy they need to turn.

In the United States, mass transit use has hit a 50-year high, thanks to rising gas prices, according to the American Public Transit Association. In the past, even when gas prices dipped, many people stayed on board their cities' buses and subways.

What You Never Learned in Driver's Ed

Although not owning a car is the best thing you can do for the atmosphere, we understand that many people do need their own vehicles because of the way that many modern cities and societies are designed. If you do need a car, don't feel guilty — and don't think that you can't help fight global warming. You can ensure your vehicle is as green as possible in several ways, from what you drive, to how you drive it, to how many passengers you bring along. And cars are improving in climate-friendly design, too, with new technologies on the horizon that will leave gasoline-powered cars in the dust.

Choosing a climate-friendly car

After the Oil Shock in the early 1970s, when oil prices skyrocketed, new government standards required passenger cars to become more fuel efficient. Unfortunately, in North America, that trend stalled: by 2006, personal vehicles were actually less fuel efficient than the fleet of 20 years earlier.

The decline was due in part to the rise of a new beast on the road: the sport utility vehicle (SUV). The SUV, whose market share has soared since its introduction, is a huge, heavy gas-guzzler in comparison to your standard passenger car. The Sightline Institute, a Seattle-based environmental think tank, has estimated that an SUV with only a driver (no additional passengers) produces about 60 percent more greenhouse gas emissions than the per-person contribution of the average air flight. Classified as a light truck (which is why it doesn't fall under tougher U.S. and Canadian fuel-efficiency standards for passenger cars), the SUV and its gas-thirsty kin — pick-up trucks and multi-passenger vans — accounted for about half of the new-car market in the U.S. through the 1990s, although this number is falling because consumers are reeling from high gas prices.

In contrast, in Europe, Japan, China, and even Australia, saving fuel is much more a way of life. In most cases, governments have legislated greater fuel efficiency (or likely will, if car manufacturers don't meet voluntary standards). But lifestyle also plays a role: Drivers in Europe, in particular, favor small vehicles that can more easily maneuver narrow streets, park in tight spots, and save on gas (which is much more expensive than it is in North America).

Table 17-1 provides some comparisons of vehicles available today.

Table 17-1 The Good, the Bad, and the Efficient Vehicles

Vehicle	*Miles per Gallon (km per Liter)*	*Rating*	*Lowdown*
Scooter or moped	60–160 (26–68)	Top environmental choice	Ideal for inner-city driving; uses less gas than a motorbike; powerful engine; storage space; two wheels; low cost, as compared to a car
Motorbike	35–55 (15–23)	Top environmental choice if it's a high fuel-economy bike	Fuel economy has wide range; two wheels; good for long-distance riding; special license needed in some countries
Small hybrid car	52 (22)	Top environmental choice	Harvests otherwise wasted power; switches between electric and gas engines, as needed; gets better mileage in the city than on the highway — the opposite of standard gas-powered cars

(continued)

Table 17-1 *(continued)*

Vehicle	*Miles per Gallon (km per Liter)*	*Rating*	*Lowdown*
Large hybrid car	32 (14)	Top environmental choice	Harvests otherwise wasted power; switches between electric and gas engines, as needed; gets better mileage in the city than on the highway
Mini car	40 (17)	Top environmental choice	Three-cylinder engine; very compact, 2-seater, highly fuel efficient, and among the safest in its rating; great for city driving and parking. Best known is the "smart" car
Small car	28 (12)	Very fuel efficient	Small size makes them good for city driving
Large car	26 (11)		More efficient on highway than in city
Vans and minivans	26 (11)	Efficient if full!	Efficient for transporting full loads of people and/or equipment
SUVs and trucks	17 (7)	Least fuel efficient	Highly useful for transporting stuff; Hummers on a downtown street are the least efficient; diesel trucks create lower emissions than trucks running on regular gas
Hummer limousine	8 (3)	Uh . . .	Highly useful if it comes with David Beckham inside; otherwise, a waste of space

Note: Miles per gallon calculated on average between city/highway driving and automatic/standard transmissions between various vehicles currently on the market.

Although choosing a car within one of the more efficient categories in Table 17-1 can help ensure that you waste less gas, fuel efficiency can vary widely within a category. Checking online at one of the following Web sites can steer you to the most fuel efficient vehicles within a particular class:

- **Australia:** www.greenvehicleguide.gov.au
- **Canada:** http://oee.nrcan.gc.ca
- **United Kingdom:** www.dft.gov.uk/actonco2 (click the Best on CO_2 Rankings link)
- **United States:** www.fueleconomy.gov

The more fuel efficient the car, the lower its emissions and the better it is for the environment. Usually, smaller cars are more fuel efficient. Ninety percent of the gas in a car is used not to move people around, but to move the car itself. Because the weight of the car is such a big factor, carrying two passengers uses about the same amount of fuel as carrying just one. So, going from one passenger to two basically doubles the efficiency of a car on a per-person basis.

Helping out the environment with a hybrid

The most efficient category of car is the hybrid. A *hybrid* is any vehicle that relies on more than one form of power. Train locomotives, for example, are diesel-electric hybrids. Although hybrids have been around for more than a century, they're just picking up steam again in the car market. The hybrid's high fuel efficiency is based on the fact that gas-powered cars waste a lot of the energy that they produce (close to 80 percent, in fact). For example, when you apply the brakes in a regular car, you actually work against the power of the engine. That power gets turned into heat that just dissipates into the air. Similarly, when you coast, the engine runs but doesn't actually help propel the car. (Think of pedaling your bike while you rocket down a very steep hill — your legs might be pumping like crazy, but gravity's moving you forward, not your legs.)

A hybrid, which has both a regular combustion engine and an electric engine under its hood, converts the energy normally wasted during braking or coasting into electricity. It then uses that electricity to help propel the car when the car needs power again. The electric motor works on its own in low-speed driving conditions (conditions in which gas engines are least efficient). The gas engine kicks in and helps out when the batteries alone can't provide enough power. So, the vehicle uses a smaller gas engine than a non-hybrid of the same size, further boosting its fuel efficiency.

In addition to having two separate engines (gas and electric), hybrids differ from conventional cars in other ways. Hybrids

- Carry a bank of batteries to store electrical power, not just the single one in a conventional car. The hybrid's electric motor can charge the batteries, as well as draw energy from them.
- Include a sophisticated onboard computer that directs the transfers of energy whenever it sees an opportunity to save gas.
- Feature innovative drive trains in some models that further increase their efficiency.
- Display in real time how much gas is used, as well as the average gas usage per mile. This information helps drivers avoid moments of rapid acceleration or speeding; the displays tell drivers just how much gas they're wasting when they put the pedal to the metal.

If you want to know what it's like to drive a hybrid, you can give it a shot through most rental companies. After you try a hybrid, you might want to buy one. (After not owning a car since 1980, Elizabeth just became the proud owner of a hybrid car.) Just be prepared for that first stop sign, when you think for a second that the engine has died. The gas engine turns off when you brake because the car doesn't need it to slow down or sit still. As soon as you need power, press the gas. Depending on the decision that the onboard computer makes, the car may move silently forward on battery power, or you may hear the sound of the car engine using gas.

Driving on a dime: Ways to use less gas

Even if you're not in the market for a new car, you can increase your own vehicle's fuel efficiency by up to 20 percent just by giving your car regular tune-ups and using fuel-efficient driving skills.

Here are some ways to drive more efficiently:

- **Turn off your engine when you stop your car for ten seconds or longer, and avoid turning on your engine before you need to.** An average newer car uses about the same amount of gas in ten seconds as it does to re-start the engine. If you turn off your vehicle whenever you're stopped for ten seconds or longer (when you're stuck in a traffic jam or pulled over — not when you're at a stop sign or a red light), you can easily save money and reduce emissions. And you also help out your vehicle.

  Extensive idling actually damages modern engines, and (contrary to popular belief) you can best warm up a cold engine in winter by driving it, not letting it idle. You can save on gas costs, as well: For every two minutes a car idles, it uses about the same amount of fuel it takes to go

1 mile. By idling, you're really going nowhere fast — especially because medical studies have linked car exhaust to asthma, allergies, heart and lung disease, and cancer.

- **Opt out of the drive-through.** Often, it takes less time to park the car and walk in than use the drive-through. With the money you have to spend on the gas you use while sitting and waiting, you could buy two sundaes rather than one. You can even join the local movement in many towns and cities to ban drive-throughs.
- **Keep tires fully inflated to the recommended pressure.** Keeping your tires pumped up increases your car's miles per gallon.
- **Remove the roof rack during seasons in which you don't use it and remove the mud flaps behind the wheels during the summer.** Believe it or not, removing these add-ons improves the aerodynamics of your car and reduces drag when driving, which increases your miles per gallon.
- **Don't stomp on the brake or the accelerator unnecessarily.** You really waste a lot of gas when you gun the engine, and when you immediately cut off the momentum you had. Try to avoid road rage moments!
- **Drive at the speed limit, rather than over it.** Fuel use, carbon dioxide emissions, and speed are directly related. According to a report by the European Transport Safety Council, simply enforcing the speed limit of 70 mph (113 km/h) in the U.K. would cut emissions by 1 million metric tons each year, and lowering that speed limit to 60 mph (97 km/h) would cut emissions by another 0.9 tons each year.
- **Take all that junk out of your trunk.** Extra weight means your car uses more gas.
- **Keep up with maintenance.** The more smoothly your car runs, the less gas it uses.
- **Run the air conditioning only when you need it.** Most cars use engine heat to warm a car, but it takes extra engine power to cool it down — enough to lower the miles to the gallon you're getting. (Of course, rolling down your windows significantly increases drag if you're on the highway, so use this tip when on shorter or inner-city trips.)
- **Drive in the highest gear possible.** In a manual-shift vehicle, driving in a high gear reduces the need to step on the gas peddle, and it also takes stress off of your engine.

All the tips in the preceding list also apply to running your boat, motorbike, scooter, dirt bike, four-wheeler, jet-ski, snowmobile, golf cart, and so on. Even though cars and trucks produce most transportation-related emissions, get into the habit of driving everything more efficiently.

Sharing a ride

According to many estimates, sharing a vehicle can do more to reduce carbon emissions than avoiding air travel (we talk about air travel in the section "Joining the Real Mile-High Club," later in this chapter). You can share a car by offering to drive other people (or hopping into someone else's car) or by owning a piece of a fleet of vehicles. Either way, you're driving less, which means you're responsible for fewer greenhouse gases entering the atmosphere.

Carpooling

Carpooling, when individuals share a ride to the same destination, takes a load off you because you don't have to drive every day. Office carpooling, when a group of people share a ride into work, is the most common kind, but countless other possibilities exist. You can carpool in almost any situation, such as driving kids to practices after school. Sharing driving duty with three other parents, dividing up who drives the children which days, can make your life three times easier. Even rock stars are helping their fans get into the act: The Dave Matthews Band sets up carpooling to and from its concerts via its Web site.

Many highways and major roads in and around cities have special lanes designated for cars that have more than one person in them — a huge benefit during rush-hour traffic. Some cities offer special parking lots where carpool members can meet up.

The Internet has made ride-sharing easy. A lot of regions have their own sites dedicated to connecting drivers with passengers (which you can find through a Web search), and some social networking sites offer ride-share applications. Through these online services, you can meet the people with whom you plan to share a ride and settle how you want to split the costs for the trip.

A car that has five riders is actually almost as efficient per person as a moped carrying one person or someone's share of a commuter train trip.

Car sharing

You can also share wheels through a *car-share* program, in which a group of people collectively own a fleet of cars as a cooperative. Each person pays for a membership and schedules his or her use of a car. Joining a car-share program can cost a lot less than renting a car, and you don't have to worry about any of the maintenance hassle of individual ownership. Even better, you often drive low-emission cars in this kind of program. Car-share programs exist around the world. To find one near you, check out `www.carsharing.net/where.html`.

Watching for upcoming car technologies

Auto shows used to be the domain of people interested in the power underneath a car's hood, hungrily eyeing prototypes that promised more horsepower in the engine. Now that manufacturers are flaunting greener models and exploring ways to get cars off oil entirely, auto shows attract more people simply looking for a practical, low-carbon car. At recent car shows, automakers featured hydrogen, electric, and even compressed-air cars — all potential alternatives to fossil-fuelled vehicles.

Using rocket science

Cars of the future may use the same technology that sends rockets into outer space — combining hydrogen and oxygen to produce electricity, with water vapor as the only emission.

Although the scientific breakthrough behind this idea of a fuel cell happened almost 200 years ago (Sir William Grove invented a gas voltaic battery in 1845), it may take a few more decades before fuel cells can run the cars on the road. Several wrinkles need to be ironed out first:

- **Cost:** Current fuel-cell technology is very expensive — about three times the cost of conventional fuels.
- **Electricity:** Producing and compressing pure hydrogen for fuel-cell cars (and other applications for fuel cells) requires electricity. If that electricity is produced by using fossil fuels, then you don't significantly reduce the emissions tied to your car.
- **Infrastructure:** Companies argue that they need support from governments to widely build hydrogen fueling stations, which these cars need to stay running. One of the first fuel-cell infrastructures is the Hydrogen Highway linking Vancouver and Whistler, British Columbia, built for the 2010 Olympics.

Cars that run a conventional internal-combustion engine by burning hydrogen, rather than (or in addition to) gas, face similar problems. Although manufacturers are road-testing the first of these hydrogen cars, they admit that these cars are probably too expensive for the average driver and that fueling stations are too few and far between.

Because hydrogen doesn't occur naturally in the pure form that these cars need, the fuel company has to produce it from a hydrogen-containing molecule, such as water, ethanol, or a fossil fuel. So, a whole wind farm could provide the energy for the process to separate the hydrogen from water (a carbon-free solution), or power from a coal-fired plant could separate the hydrogen from methane. Hydrogen isn't necessarily a clean, green energy source — but it can be.

Capturing hydrogen from water

Hydrogen has the potential to be an ideal fuel because people can find it in something the Earth has a lot of — water, which is composed of hydrogen and oxygen. To separate the hydrogen and oxygen atoms from each other, two electric currents are put through the water: one negative and one positive. The charges of these currents cause the chemical bonds between the hydrogen and oxygen atoms to separate.

Every atom in chemistry has a charge. Hydrogen is positive and oxygen is negative. They act just like two magnets that pull together because they're attracted to the opposite electrical charges, which are stronger than the bond between each other. The hydrogen and oxygen each bubble up as a gas, allowing pipes to capture the hydrogen and compress it into a fuel cell. Many developments are underway to find more efficient ways to make hydrogen fuel cells. Currently, the process is too energy-intensive to be widely used.

Currently, hydrogen fuel cells aren't employed on the road, although some forklifts use the technology. Public transit may take to the hydrogen highway first because those vehicles refuel in central locations. Many cities have already acquired hydrogen buses or are testing the technology. For private cars, hydrogen technologies await their commercial breakthrough and, in the meantime, remain controversial because of the amount of energy required to produce fuel cells in the first place. In the long term, civilization is likely to use fuel cells in a wide range of applications beyond vehicles, including powering buildings.

Electrifying vehicles

The electric car runs by using rechargeable batteries, so it doesn't need any fuel on board at all. It was invented in the mid-1800s, and by the beginning of the 1900s, one-third of all cars were electric. However, they rapidly lost popularity after gas-powered cars adopted electric starters (no more hand-cranking) and the motoring public started ranging farther afield — to places where you couldn't recharge an electric car.

The 1973 Oil Shock renewed interest in the electric car, and so did growing concerns about air pollution and greenhouse gases. You may remember in the mid-1990s that everyone seemed to be talking about the electric car as the car of the future. So, what happened to those great plans? Many argue that car and oil companies purposely killed plans for electric cars because of the potential threat they represented to the booming oil industry. The documentary *Who Killed the Electric Car?* goes into this connection in great detail. Electric car technology still exists, but it needs to find its market.

Like with the hydrogen fuel-cell car, the electric car is only as green as its electricity. If the electricity originates from coal-fuelled generators, for example, these cars are still responsible for a great deal of greenhouse gas emissions.

Traveling on air

Air power could be the most revolutionary car technology if someone can bring it into mass production. Because the fuel is air, only air comes out of the tailpipe. Electricity pumps air into the car's tanks until it reaches a level of pressure high enough to run the engine, instead of combusting gas. (So, the source of the electricity is important, just like for hydrogen and electric cars, which we discuss in the preceding sections.) The Mexican government has already agreed to buy 40,000 compressed-air cars to use as taxis in Mexico City, in a bid to reduce its infamous air pollution.

Joining the Real Mile-High Club

Plane travel is extremely polluting: Scientists believe that greenhouse gases released by aircraft at higher elevations actually do more damage than emissions released at the earth's surface. Several factors — empty seats, numerous connections (and therefore numerous take-offs and landings) in one flight, and even the weather — make air travel the most variable and carbon-intensive means of transportation in terms of greenhouse gases produced per passenger, per mile. So, when you can't avoid flying, make lower-carbon choices about how to do it.

Choosing when to fly

The most energy-intensive part of the flight is take-off, so a non-stop flight is best — only one take-off. For the same reason, short flights are less efficient than long flights. You get all the way up there only to hang out for half an hour before coming back down! You can most easily replace a short-haul plane trip with a bus or train ride. When you factor in airport check-in time and the fact that many train stations are in the heart of cities' downtowns, you probably don't save any time at all by flying.

Traveling guilt-free by using carbon offsets

Carbon offsetting is a way of sort of undoing the emissions from your air travel by financially sponsoring an activity that reduces emissions elsewhere by the same amount. So many people have decided to carbon offset that a lot of organizations, both for-profit and non-profit, have sprung up to serve the market.

Some examples of offset projects that you can put money into include the following:

- Retrofitting homes in low-income communities with energy-efficient light bulbs
- Installing solar panels in a community that would otherwise get its electricity from a coal-fired plant
- Investing in a small wind-power company so that it can feed into the electricity grid

Carbon-offsetting organizations are usually Web-based. Most sites offer an online carbon calculator that helps you determine just how much carbon dioxide your trip will release into the atmosphere. Figure 17-1 shows you what a typical online carbon calculator looks like.

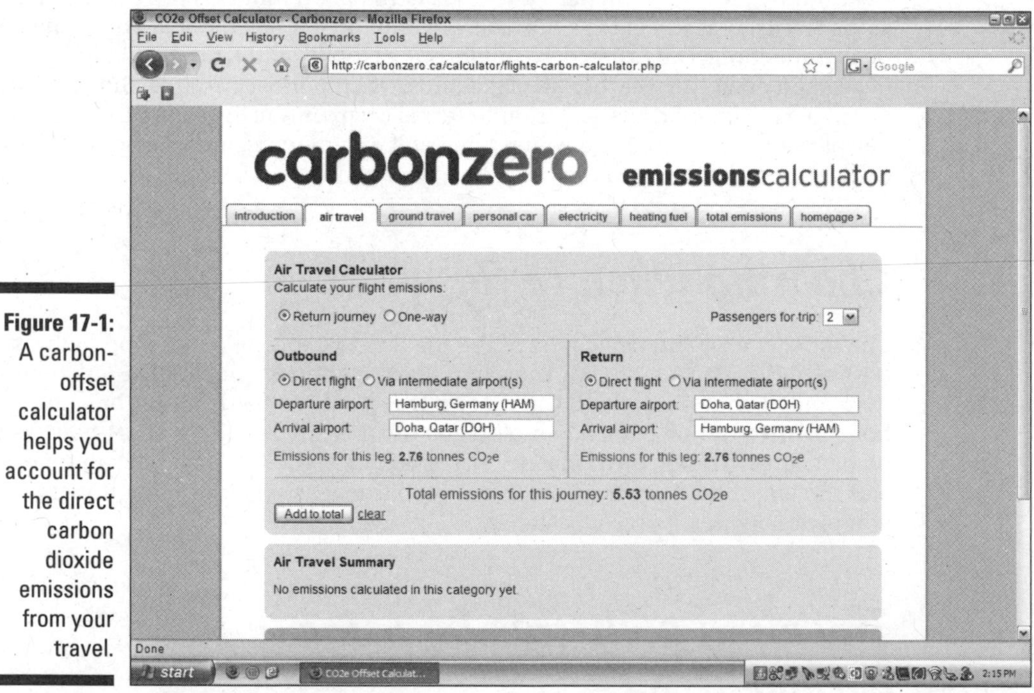

Figure 17-1: A carbon-offset calculator helps you account for the direct carbon dioxide emissions from your travel.

Carbonzero.ca

The carbon-offsetting option to plant trees is not as effective as most other offset projects. Planting trees is effective, but in terms of global warming, those trees won't start to soak up significant amounts of carbon dioxide for 25 to 30 years, and trees are vulnerable to carbon loss because of forest fires. For quick results — which you want to focus on because of the pressing urgency of climate change — choose energy projects that implement either efficient infrastructure or technology, or that involve a switch to a renewable energy source.

Although they tend to be based in the U.K. or the U.S., most carbon-offsetting programs serve people in any country and can calculate the price in various currencies. The programs tend to calculate your emissions differently because each program gives differing weights to various factors. We list the most credible and reliable sites in Table 17-2.

Table 17-2 Joining a Carbon-Offset Community

Web Site	*Based In*	*Leading Projects*	*Cost to Fly Roundtrip New York to London (in U.S. Dollars)*
Atmosfair (`www.atmosfair.de`)	Germany	Solar energy in India; local energy-saving at German schools; bio-energy in Thailand	$160
Carbon Balanced (`www.carbonbalanced.org`)	U.K.	Rainforest reforestation worldwide	$72
CarbonCounter.org	U.S.	Truck-stop electrification, wind farms, and energy efficiency in the U.S.; forest conservation in Ecuador	$35
CarbonNeutral.com	U.K.	Hydropower in Bulgaria; solar lighting in India; methane control in the U.S.	$19–$35

(continued)

Table 17-2 *(continued)*

Web Site	***Based In***	***Leading Projects***	***Cost to Fly Roundtrip New York to London (in U.S. Dollars)***
CarbonZero.ca	Canada	Wind power and energy retrofits in Canada	\$72
ClimateCare.org	U.K.	Energy-efficient lighting around the world; wind energy and bio-energy in India	\$27
co2balance.com	U.S.	Efficient lighting in Kenya; creating forested areas in the U.S.	\$33–\$72
myclimate (`www.my-climate.com`)	U.S.	Renewable energy and energy efficiency in developing countries	\$90
NativeEnergy.com	U.S.	Local wind farm developments in the U.S.	\$36

Note: Calculated emissions in terms of metric tons are different for each service provider; the average amount of CO_2 emitted by a direct, roundtrip London–NYC flight is 2.4 metric tons. Each service charges varying amounts per metric ton offset.

If you know a frequent flier, you can purchase offset gift certificates. As Table 17-2 shows, you can get most offsets for a reasonable price.

Chapter 18

Making a Difference at Home and Work

In This Chapter

- Cutting back on your home's greenhouse gas emissions
- Building environmentally friendly homes
- Reducing, reusing, and recycling your way to a cleaner atmosphere
- Choosing carbon-friendly food
- Shopping green
- Making positive changes at work

You've probably heard the saying, "Think globally, act locally." Well, it doesn't get any more local than your home, where you can make plenty of changes that help reduce your carbon footprint. Don't worry — we're not going to suggest that you give up your worldly possessions and go live in a log cabin (although if you do, be sure that the logs are from sustainable forestry practices). In fact, some of the biggest changes you can make, environmentally speaking, are some of the smallest.

In this chapter, we take a look at how you can give your life a green makeover.

Home, Carbon-Free Home

By making changes right at home, you can help do your part to cool down the planet. Reducing energy use is the name of the game because most energy (even electricity) involves greenhouse gas production, and a lot of great practices and technologies can help. (Check out Chapter 6 to see how homes contribute to greenhouse gas emissions.)

To kick off your own adventure in greater energy efficiency, set up a home energy audit, in which a trained expert goes through your home to tailor advice to your needs. In some countries, the government supports energy audits with rebates or other incentives. Either way, the investment in expert advice is well worth it — if you follow the audit advice, you can save much more money than you paid for the audit in a short time.

You can best reduce home-related emissions *immediately* by using less energy. Not only does conserving energy cost you nothing, it actually saves you money on your utility bills while saving the planet. "Waste not" is the core principle: Don't use it if you don't need it. The following sections offer conservation tips to help you cut back on your home's energy consumption.

Heating and cooling

Taken together, heating and cooling constitute the two largest uses of energy in your home — so they're also the places you can potentially save the most. Turning up your air conditioning and lowering your heat by just 3.5 degrees Fahrenheit (2 degrees Celsius) saves more than a metric ton of carbon dioxide in one year.

Automated temperature control can make your home comfortable while reducing energy use. Programmable thermostats enable you to preset temperature levels, and getting one for your home is about the best investment you can make to help save on cooling and heating costs. You set the temperatures for various periods during the day and night, as well as for the week and weekends. Not only do you not need to remember to change the settings, the program can make sure the temperature is comfortable by the time you get up in the morning or return to the house at the end of the day.

Heating

You can start saving money and reducing greenhouse gas emissions right away by getting a more efficient heating unit. If you can access natural gas, a high efficiency gas furnace makes a good investment. If you're stuck on home heating oil, you can still make a major leap forward in energy efficiency by investing in a modern and more efficient unit. Better yet, switch your system to propane, a heat pump, or even a high-efficiency, wood-pellet system. Have an evaluator come in to assess your current heating system and see what changes you can make to your home.

Day-to-day habits make a big difference, no matter what heating source you use. Keep the heat turned down when you're not home. (But remember to keep your home warm enough to prevent frozen pipes.) In fact, even when you're home, try a lower temperature than the standard 72 degrees Fahrenheit (20 degrees Celsius). Just throw on some warm and cozy clothes — over your natural-fiber long underwear, natch — instead of hanging out in your skivvies with the heat going full blast. That change drops your winter heating bill instantly. At night, you can drop the temperature even more. Your body cools off by about 3.5 degrees Fahrenheit (2 degrees Celsius) when you're sleeping. So, you can keep your house 3.5 degrees cooler at night, and you won't even notice while you dream about sugarplums and wind turbines. For even greater savings, turn the heat way down at night, invest in flannel sheets, and pile on another blanket or two.

Cooling

Increasing reliance on air conditioning has completely changed the electricity sector. Air conditioning runs on electrical power. It removes heat from the air in the house, dissipates the heat outside, and circulates the cooled air inside the house. Winter used to be the time of greatest demand for electricity, but that's now shifted to summer.

Cooling systems don't have to run on electricity or consume it greedily, however. Here are a few low-tech and low-emission options that can help you keep your cool:

- **Opt for natural ventilation.** *Natural ventilation* involves the movement of air through and within a house. A well-designed home allows for cross breezes that keep air circulating. You can help the breeze out by using electric-powered fans, which require far less electricity than air conditioners. By moving the air around, they make you feel cooler by about 7 degrees Fahrenheit (4 degrees Celsius). You can also use fans in the winter to recycle the warm air near the ceiling and keep the house more comfortable.
- **Plant deciduous trees to provide shade.** Trees can reduce air-conditioning needs by 25 percent. (And, as an added bonus, trees also absorb carbon dioxide.)

  Some state utilities in California actually plant trees around private homes for free to help reduce electricity demand. Many cities now have urban forestry programs to plant street trees at low or no cost to the homeowners, and some not-for-profit groups offer advice and low-cost trees for planting in back or side yards.
- **Keep your curtains closed during the day and open at night.** By following this curtain suggestion, you block out the day's sunlight (while you are out of the house) and benefit from natural nighttime cooling.
- **Consider a dehumidifier if you live in a humid climate.** Dryer air feels much cooler.
- **Plant a roof garden, if you can.** The natural insulation that a roof garden provides offers another way to reduce air-conditioning use by 25 percent. You can even use that garden to grow your own veggies.
- **Paint your roof white.** White roofs substantially reduce the need for air conditioning because it reflects rather than absorbs the sunlight and heat.

Pump it up: Heating and cooling from the ground

Ground source heat pumps are an effective way to help heat and cool your home while seriously reducing your carbon footprint. Geothermal heating and cooling uses the steady temperature of the ground to heat and cool your house. These pumps draw their heat (or cold) from under the earth, which holds a steady temperature year-round below 8 feet. The air that a heat pump pulls from the ground isn't that warm, but it is warmer than the outside air in the winter and cooler in the summer. The steady temperature of the ground is used as a source to pump heat either into or out of your home, working the same way that your refrigerator works. The technology is called a *compression cycle,* but only the refrigerant, not air, is compressed. You can even capture and redirect the excess hot air produced from heating and cooling to help heat your water.

Ground source heat pumps aren't cheap, unfortunately, often costing about twice the price of a comparable furnace. That initial cost pays off in the long run, however; you get energy savings of anywhere from 30 to 60 percent. Because heat pump technology is complex, these systems need to be installed by qualified contractors.

If you feel like you can't get by without air conditioning, avoid central systems, which cool your entire home. Opt for room units and buy the most efficient ones possible. By using room units, you can cool just the room you're in, instead of needlessly cooling every room, regardless of whether anyone's in them. And you don't need to cool an empty house. Leave your air conditioner off, or on a warmer setting, if you're not home.

Drafts

Any cracks and crevices in your home increase its greenhouse gas emissions by reducing its efficiency. If hot air is streaming from your home in the winter and cool air is escaping in the summer, your heating and cooling systems have to work overtime.

Check for air leaks in your home on a windy day by lighting an incense stick (any scent will do) and holding it up to any spot that's a possible air path to the outside, such as an air duct, window, electrical outlet, or plumbing fixture. If the sweet-smelling smoke blows horizontally, you have a leak that you need to seal.

Sealing up your home

Here are a few tips to keep your home from losing any warm or cool air:

- Look for entrance doors, cat doors, and even mail slots that don't close all the way or that don't totally seal. Getting those fixed up is cost-efficient and can really help conserve energy.

- Open and close the door right away when you're going into and out of the building.
- Invest in good curtains and keep them snugly closed at night in the winter to keep in the warmth.
- Check the weather stripping around doors and other openings every fall, replacing or upgrading it, as needed.
- Check the chimney. If you have a fireplace in your home, the chimney is a great escape route for heat. Seal off the chimney completely if you don't use the fireplace. If you do use it, be sure to close the damper when it's not in use. Even better, install a fireplace insert or wood stove in the opening. Either option gives you a dramatic improvement in energy efficiency because most of the heat that a conventional fireplace produces goes straight up the chimney.

Getting the most from your windows

Many homes lose efficiency through their windows. Replacing old windows with new and more efficient ones may be worth doing, but if you need to prioritize, you're probably better off improving your home's insulation and the efficiency of your furnace first. (An audit is so important because it informs you where your efforts can make the biggest difference.)

When shopping for energy-efficient windows, look for these characteristics:

- **Double paned:** Two panes of glass, separated by a gap, increases a window's insulation value while providing the same clarity.
- **Gas-filled:** Gas between the panes of glass acts as extra insulation.
- **Self-sealing:** Soft rubber, which runs along the edges of the window, acts as a seal when the window is shut so that absolutely no air sneaks through.
- **Energy label:** Often, windows are marked with an Energy Star or Energy Saving label to show that they meet efficiency standards. (Check out the "Appliances and electronics" section, later in this chapter, for more about these labels.)

You don't have to replace your windows to make them more efficient. You can seal drafty windows simply by caulking them. Another winter sealing option for cold climates involves covering the inside of the window with clear plastic. You can pick a kit up at any hardware store, and all you need is a hair-dryer (hopefully, an energy-efficient one) to finish off the application. Or, for a higher cost (that's still less than replacing the window), you can use removable indoor or outdoor storm windows.

Insulation

You can conserve energy by improving your home's insulation. *Insulation* is material that slows the escape of heat from your house in winter and keeps the heat out of your house in summer. It comes in many different forms — the most common include fiberglass *batts* (sheets), foam board, and cellulose or fiberglass loose fill. Insulation is usually installed during construction. Traditionally, it went only into the attic, but more modern houses are insulated between the walls when they're built. Insulation can especially make an energy difference if you live in a climate that has extreme temperatures, hot or cold. With enough insulation, a building may not even need a furnace.

The U.S. Home Energy Guide suggests that you check the following spots to ensure they're insulated, listed in order of importance:

- Attic
- Ceilings under unheated spaces
- Exterior and basement walls
- Floors over unheated spaces
- Crawl spaces

Insulation is measured in *R-values* (called RSI-values in Canada), such as R-40 or R-25. R-values go up to 60, with better insulating ability the higher the number. The R-value recommended for your house depends on what kind of climate you live in. You can probably best determine your insulation needs by having an energy audit, but in the United States, you can use the Zip-Code Insulation Program (`www.ornl.gov/~roofs/Zip/ZipHome.html`). This Web site determines the most practical level of insulation for your home, based on where you live and your type of heating.

Depending on where you need it, adding or upgrading insulation can be a complicated job requiring specialized equipment and skills. You need to take into account vapor barriers and other considerations. Gaps reduce the overall effectiveness of the insulation, and so does packing insulation in too tightly. Also, you need to make sure that your home is properly *air sealed* before you insulate — ensuring there are no cracks or spaces in your exterior walls for drafts to get through. Finally, some types of insulation in older houses contain materials that are potentially toxic if disturbed. If you decide to tackle the work yourself, be sure to do the required research and take the necessary precautions.

Appliances and electronics

Out with the old, in with the new — new, energy-efficient wonders, that is. Household appliances and other products have become incredibly energy efficient over the last several years. In many cases, the newer an appliance is, the less fossil-fuel consumption it triggers. Making the right decisions today doesn't just lead to immediate savings, it can save energy for the next 10 or 20 years. Many national governments have set up national energy standards for appliances. For many other products, federal standards don't exist, but ENERGY STAR can help.

ENERGY STAR was introduced by the Environmental Protection Agency (EPA) in 1992 as a voluntary, market-based partnership to reduce greenhouse gas emissions through superior energy efficiency. Today the ENERGY STAR label is on more than 50 different kinds of products, on new homes, and on commercial and industrial buildings. Products, homes, and buildings earn the right to display the ENERGY STAR by meeting strict energy-efficiency criteria set by the EPA.

The savings have been tremendous. In 2007 alone, Americans, with the help of ENERGY STAR, saved more than $16 billion on their utility bills while reducing greenhouse gas emissions equivalent to those from 27 million vehicles. For more information on ENERGY STAR, go to `energystar.gov`.

Figure 18-1 shows the label to look for in the U.S. when shopping for energy-efficient appliances. Hunt for ENERGY STAR in the U.S., Australia, and Canada, and for Energy Saving (of the Energy Saving Trust) for the U.K. Both labels are bright blue and both detail their energy-use in relation to a regular, non-energy-efficient appliance, making it easy to see the savings before you even make a purchase.

Figure 18-1: Look for the ENERGY STAR label.

Environmental Protection Agency

Making the right choices when replacing major appliances can make your energy bill drop immediately.

REMEMBER

Always check the energy rating. Just because an appliance is new doesn't mean it's energy efficient; some new appliances are still energy hogs.

Fridge and freezer

Refrigerators can be true energy hogs. Happily, Energy Star and Energy Savings fridges use about 40 percent less energy than any model made before 2001. Here are a few suggestions for making your fridge even more efficient:

- Locate it away from warm appliances, such as the oven, stove, and dishwasher.
- Make sure you open the fridge door only when needed and seal it tight when you close it. (You can test your fridge's seal by closing the door on a piece of paper — if the paper stays, the door is sealed.)
- Set the temperature in your freezer at 0 degrees Fahrenheit (–17 degrees Celsius) and the fridge at about 35.5 to 40 degrees Fahrenheit (2 to 4.5 degrees Celsius) to keep things cool.

REMEMBER

When you buy your new energy-efficient fridge, don't keep the old one in the basement. Get it properly recycled, which safely removes its cooling liquids to avoid releasing ozone-depleting substances.

Dishwasher

You may be thinking that a dishwasher is a wasteful extravagance, but believe it or not, you can use a dishwasher without guilt. A study by the University of Bonn, in Germany, found that washing a full load in the most energy-efficient dishwasher, such as those with the Energy Star and Energy Savings labels, uses half the energy (from hot water use) and one-sixth the water that washing the dishes by hand does. Ensure that you choose the air-dry option, avoid the pre-rinse or rinse-hold options, and sit back while the machine does your work, confident that you're doing the atmosphere a favor!

Still, we include a lot of caveats before endorsing a dishwasher! Overall, if you wash a sink full of dishes by plugging and filling the sink (so that you don't waste water), hand-washing saves energy. If you must have a dishwasher, buy the most energy-efficient model and don't run it until it's full. Remember not to run your dishwasher until after peak energy demand — wait until the last thing at night to switch it on. *Peak times* are when the electrical generating units in your area are straining to meet demand. Shifting use to off-peak times allows the generating station to run more efficiently and, if coal-burning, burn less.

Turn off "instant on"

Many appliances, such as televisions, video game systems, and portable stereos, offer an "instant on" feature, coming to life the moment you press the power button. What "instant on" really means, however, is "never off." These devices drain a constant stream of electricity so that they can leap into action when you need them. Although the trickle of power that they use may seem trivial, an average household can reduce electricity consumption by as much as 15 percent by sidestepping this feature. Plug your "instant on" appliances into a power strip, and when they're not in use, turn the power strip off. (An even simpler option? Just unplug the appliances when you're not using them.)

Washing machine and dryer

The most common washing machine model is the top-loader. Unfortunately, these models aren't that efficient, using more water to wash your clothes than front-loading washers. Top-loading machines are also less efficient at spinning, meaning that they make your dryer work harder. Although front-loading machines are more expensive, the investment can pay off in energy savings, not to mention the reduced environmental impact.

To be truly sure your washer is green, buy ENERGY STAR or Energy Savings washers, which use about 50 percent less energy than other models, and up to 22 fewer gallons of water per full load.

Unfortunately, you can find very few eco-savvy dryers. Most dryers use about the same amount of energy. You can save some energy by buying a model that has a moisture sensor option; the dryer automatically turns off after the clothes have dried. The best option, however, is low-tech: Try a clothesline, which is quite "eco-chic" these days.

Clotheslines are banned in some communities because people think they're unsightly. If you can't hang your clothes out to dry in your community, you can call your elected representatives to complain.

Multimedia electronics

These days, just about any electronic device comes in an energy-efficient version. Here are some bigger electronics that are available with low-energy ratings:

- **Televisions:** LCD screens are the most energy efficient; the least are the plasma screens larger than 50 inches. (Plus flat-screen TVs use nitrogen trifluoride [NF3], a greenhouse gas with warming properties 17,000 times more powerful than carbon dioxide. The popularity of flat screens has created a new climate risk. NF3 was not covered in the Kyoto Protocol.)

- **Computers:** You can find energy-efficient models in desktops and laptops, but laptops take far less power, using only a tenth of the electricity drawn by a desktop.
- **Sound systems:** Including amps, speakers, and large stereo systems.

More careful computer use can save a huge amount of energy at home or at the office. First, turn your computer off — much of the energy that computers use is wasted because you leave them on at night, over weekends, and even for extended periods of inactivity during the day. If you must leave your computer on, at least turn off the monitor, which consumes a big chunk of the energy it uses.

Lighting

The light bulb you grew up with, the incandescent bulb, is soon to be a thing of the past. This energy-inefficient artifact is being phased out all over the world, in favor of energy-saving alternatives. The U.S. government has committed to phasing out incandescents over the next 12 years; the Australian government plans to phase out the bulbs by 2010.

The most common energy-saving light source is the compact fluorescent light bulb. It uses one-quarter of the energy that a regular bulb does to produce the same amount of light and can last up to ten years. Over its lifetime, assuming the electricity it uses comes from a coal-fired plant, one bulb prevents (literally) a metric ton of carbon from entering the atmosphere.

These compact fluorescent bulbs contain a trace amount of mercury. At the end of their useful life, you can't just throw them in the garbage, you have to handle them as hazardous waste to avoid the minute amount of mercury in the bulb escaping into the landfill. (Each region has different regulations on how to handle hazardous waste — check with your local government for more information.)

The next generation of energy-efficient lighting will be LEDs (light-emitting diodes). These lights draw very low power and also last a long time, without the disposal problems of compact fluorescents. The technology is advancing rapidly, and when prices fall, people can start using them to light their homes. Many homeowners have already invested in LEDs in the form of low-energy-consuming Christmas lights.

Keep an incandescent light bulb in a box in the attic so that you can show your grandchildren what a light bulb used to look like.

Energy efficiency near you

Your national government probably offers some great online resources for checking up on home energy-saving tips, including the latest energy-efficient appliances and technologies available. If your country isn't listed below, check the Web site of your national department of energy.

- **Australia:** Energy Rating (www.energyrating.gov.au)
- **Canada:** Office of Energy Efficiency (www.oee.nrcan.gc.ca)
- **United Kingdom:** Energy Saving Trust (www.energysavingtrust.org.uk)
- **United States:** U.S. Department of Energy (www.energy.gov/forconsumers.htm)

Warm waters

According to Australia's Department of Energy, an average house using an electric water heater produces about 4 metric tons of greenhouse gas annually; a natural gas heater produces 1.5 metric tons. You can cut back on how much energy your water heater uses by making some simple fixes:

- **Wash laundry in cold water.** More and more detergent brands come in a "cold water" version.
- **Turn your water heater thermostat down.** Set the thermostat on your water heater a little lower; you don't really need your taps to go to "scalding," do you?
- **Insulate your water heater.** Invest in an insulating jacket for your hot water tank to save on heating costs; the jacket prevents heat intended to heat your water from escaping. Check with the equipment or fuel supplier for options.
- **Install a low-flow showerhead.** Many areas offer showerhead exchange programs.

If you want to make bigger changes, consider a solar water heater. You, as an average homeowner, can get this technology, and although it's expensive initially, it saves you money over the lifetime of the unit. The sun's warmth preheats your water and vastly reduces the amount of fossil fuel energy you need to use to take a steamy shower. In fact, depending on your hot water consumption and the local climate, solar water heaters can provide about 60 percent of your annual water heating needs. Most solar water heaters need an auxiliary heater (natural gas or electric) to ensure that you can meet your hot water needs when it's very cloudy or the days are very short.

Tankless (also known as on-demand) water heaters are another energy-efficient option. Instead of storing hot water (and wasting energy keeping it hot until it's needed), a tankless heater works only when you turn on the hot water. This kind of water heater is much more energy efficient than the conventional storage heater and can provide savings of up to 75 percent on your water heating bill. They work well in tandem with a solar system, acting as a back-up for situations when your solar-powered heater doesn't have enough juice stored to heat up your home's water.

Green Developments: Building or Renovating

Whether you're starting from scratch or renovating, major construction offers you the chance to really make your home energy efficient and carbon friendly. Keep the following elements in mind when working towards a low-energy and climate-friendly home (and check out *Green Building & Remodeling For Dummies,* by Eric Corey Freed [Wiley] for a whole book's worth of information on this subject):

- **Hire an eco-friendly architect.** If you're planning to use an architect, look for one who's LEED accredited. *LEED* stands for Leadership in Energy and Environmental Design, and is a system launched by the U.S. Green Building Council, which sets standards for buildings' energy efficiency and environmental stewardship.
- **Install automated systems.** Many new technologies help you to reduce your energy use. We rave about programmable thermostats in the section "Home, Carbon-Free Home," earlier in this chapter. Another system allows you to turn off every light in the house with the push of one button on your way out the door.
- **Be sun-smart.** Build the longest side of your house facing south (or north, if you're down under), and include well-insulated walls on the opposite side to capture and store heat in the winter when the sun is low in the sky. Having a lot of windows on the south side of your house helps, too. In the summer, this orientation actually helps keep your house cool because walls facing east and west are less exposed to the strong heat of the rising and setting sun. But for summer protection, you need a roof overhang on the south exposure, and the south windows need shades. (We talk about using the sun's energy to heat your home in Chapter 13.)
- **Plant trees.** Keep or plant as many trees as possible on your property. Trees *transpire* (release water into the atmosphere), which has a cooling effect, adding to the benefits of their shade.

- **Landscape smartly.** Manicured, green-carpet lawns are among the most wasteful practices of modern civilization. The water that is used to keep it growing and green, and the mowers that are used to cut it back down, both require a lot of energy. Consider alternatives, such as a clover lawn; it grows only an inch high and stays green all year. You can also reduce your lawn's energy consumption by relying on the rain or collecting rainwater from rain troughs on your roof.

If you want your lawn to be extra environmentally friendly, keep it trim with a push mower, rather than the electrical or gas-powered alternatives. You can also avoid using fertilizers on your lawn — the production of fertilizers releases nitrous oxide, a greenhouse gas over 300 times as heat-trapping as carbon dioxide.

- **Go underground.** Build part of your house underground. Whether it's for a multi-use basement or the main floor, the ground can help moderate temperatures.
- **Investigate alternative building materials.** People are discovering new ways of building or going back to old ways that are far friendlier to the environment. Forests are one of the world's key carbon sinks, so instead of using wood to frame your house, think about insulated concrete forms (ICFs). They make a house 30 to 50 percent more energy efficient and save hundreds of trees. But if you're worried about the impact of concrete, which takes huge amounts of energy to produce, you might go even further and consider building with walls of straw bale or (in hot, dry climates) rammed earth.
- **Lighten up on materials.** Kitchen "must-haves" such as granite countertops come at a huge energy cost to quarry, cut, and haul. But newly fashionable concrete countertops might be even worse because concrete is an energy hog, taking lots of power to be made. Look instead for funky alternatives, such as recycled glass or sorghum-fiber laminate. Rather than using hardwood for floors, people are rediscovering old materials such as cork, linoleum, and natural-fiber carpeting, as well as cool new floor materials such as bamboo.

If you do use wood, ensure that it's Forest Stewardship Council certified, which means the forests are grown and harvested without soil damage or clear-cutting. (Refer to Chapter 14 for more about sustainable forestry practices.) And keep in mind that a *veneer* (thin layer) of hardwood over environmentally-friendly plywood is a much greener option than a solid piece.

- **Recycle and reuse.** Most home building materials are recyclable or reusable. When renovating, be sure to save salvageable materials. If you can't reuse them, you can often find collection programs in cities that take materials such as doors and windows. Building with recycled or reused materials prevents energy from being used to produce the same thing new from raw materials. You can even buy recycled paint now.

If your city has a Habitat for Humanity chapter, they probably run a salvaged materials resale store. One clever architect managed to build a fabulous little "scrap yard" house in Kansas for only $50,000, using all recycled and reused materials.

If you're really committed to building a home that isn't a drain on energy systems, you might want to consider building a *zero-energy home* — one that isn't just highly energy efficient, but produces energy to feed into the grid. The goal of these homes is to produce as much power as they use, or even more. Governments in Canada, the U.K., and the U.S. are providing support for new techniques to create zero-energy homes, which are already a reality. One house built in Colorado by Habitat for Humanity has already met the zero-energy standard over a year of operation.

Powerful Changes: Renewable Energy

Cutting back on energy consumption in your home is great, but chances are very good that the energy you're still using isn't renewable and is producing greenhouse gases. Although people need to encourage their governments to explore sustainable energy sources (which we discuss in Chapter 13), you don't need to wait to use cleaner power.

You can benefit from renewable energy in your home today:

- **Generate the energy yourself.** How much generating your own energy costs depends on where you live and whether your government offers incentives for retrofitting your home. Solar, wind, and geothermal heating and cooling technologies are becoming more widely available for home use. Currently, these costly units take a long time to realize any energy savings, but their prices may drop with higher production and technological advances.
- **Buy renewable energy directly.** You can bring energy straight to your home from an independent power generator, as opposed to the general service provider for your region. Your energy is still delivered through the same electricity grid, but you're bringing renewable energy into the grid to replace non-renewable energy sources. You get total independence from fossil fuels, but you can use this option only if you have a clean energy services provider in your area.

- **Buy renewable energy indirectly.** Renewable energy from clean, emissions-free sources (such as wind power and low-impact water power) often costs more to produce than other alternatives (see Chapter 13). You can opt to pay a premium for your power to cover the cost of that sustainable energy. Your household energy consumption is still metered, but your payment goes to supporting renewable projects only, rather than the mix of generation sources that feeds into the grid. It's essentially the same as carbon offsetting (which we discuss in Chapter 17), but for your home.

Here's a list of Web sites where you can find information about green power in your area and how to purchase it, either directly or indirectly:

- **GreenPower (Australia):** This site links you to renewable energy producers near you. (www.greenpower.gov.au)
- **UKGreenPower (United Kingdom):** This program allows you to type in your postcode and see what options are available in your area. (www.ukgreenpower.co.uk)
- **The Green Power Network (United States):** This U.S. Department of Energy Web site outlines what's available for your state. (www.eere.energy.gov/greenpower)
- **Pollution Probe (Canada):** This non-governmental organization has a consumer guide to the green energy market in Canada that gives options by province and territory. (www.pollutionprobe.org)

Home, home in the hill

Fewer houses are greener — literally — than Dr. Bill Lishman's. His home outside of Blackstock, Ontario, Canada, is a grassy hill, covered with lush gardens. Dr. Lishman designed his home to be underground, using the earth as insulation. This alone makes the house just about as energy-efficient as you can get, but Dr. Lishman didn't stop there. The entire house is a testament to efficiency, using as little energy as possible.

Although underground, the house's interior is remarkably bright. Large sky lights in every room and white curved walls allow the light to reflect throughout the house, eliminating the need for artificial lighting (until nighttime, that is). Two sunrooms peak out of the hill and act like greenhouses during the day, growing warm in the sun. Air ducts carry the warm air from the sunrooms into the rest of the house.

For photos of Lishman's home and a detailed report of its construction, visit www.williamlishman.com/underground.htm.

Cutting Back on Waste

Modern civilization throws out too much stuff, and that waste is affecting the climate. The U.S. produces the equivalent of 4.6 pounds (2.1 kilograms) of waste per person every day. That number is a little lower in the U.K. — about 3.1 pounds (1.4 kilograms) per day — but it's risen 9 percent in nine years. That garbage sits in landfill sites, producing methane gas, one of the most serious greenhouse gases. (Refer to Chapter 2 for more about methane.) Add to that the fuel burned in transporting trash to the site and the energy exhausted to create that unwanted stuff, and humanity has a real problem on its hands. Happily, people have the power to remedy the situation by making some adjustments to their lives.

Some communities offer limited recycling or composting options. If yours is one of those, try calling your local government and asking if they have any plans to expand these programs. Write a letter to your mayor or the editor of your local paper. If you don't demand action, who will?

Producing less garbage

Aim to produce zero garbage. It shouldn't be too hard: You can recycle or compost about 90 percent of what normally gets tossed in the trash. And sadly, a lot of what people do throw out they didn't need in the first place. Did you know 25 percent of all food that U.S. households buy gets thrown out?

You can cut back on non-recyclable, non-organic waste in the following ways:

- **Plan meals.** When you know exactly what you're going to eat for the week, you're unlikely to buy more than you need. Avoiding food-related excess cuts down on both food waste and garbage, reducing greenhouse gas emissions.
- **Reuse containers.** Avoid disposable plastic bags for your lunches and leftovers, opting instead for sealable containers that you can use over and over. (Be careful, though, to use non-plastic containers so you don't get nasty chemicals leaching into your food.) Not using disposable bags cuts back on fossil fuels used to make that plastic, saves the energy used (and emissions created) from making the bags, and reduces emissions by producing less waste. As an added bonus, you save money.
- **Avoid individually packaged products.** A lot of products now come prepackaged for individual use. Sure, they're convenient, but they're not worth it in the long run. You pay more and create a whole lot more waste. Instead of going for the individual packages, buy in larger quantities and create your own individual portions in reusable containers.

- **Buy in bulk.** At many natural food stores, you can buy not only food but washing detergents and shampoos in bulk. Save and reuse the containers for bulk purchases.
- **Avoid disposable bottles and cans.** When possible, buy products in refillable containers — but be sure to get them refilled! (Check out Chapter 5 to see how much energy goes into making aluminum cans.)
- **Avoid ordering take-out.** Driving your meal to your home, the waste from the disposable containers, and so on — the problems with take-out are many. If you do order take-out, ask what kind of packaging the restaurant uses to find out whether you can recycle that packaging.
- **Try not to buy on impulse.** All it takes is a pause to think about why you're buying something — more often than not, you don't really need the product, and it just ends up in the garbage, contributing to the global warming problem.

Recycling

Recycling saves energy: It takes much less energy to melt down an aluminum can to make another can than to process the raw materials to make a can from scratch.

Most materials are recyclable, but what you can recycle depends on where you live. You can call your city or town, or visit its Web site, to find out what you can recycle in your area, as well as what gets picked up on the curb versus what you have to drive to the depot yourself. Here are the materials most commonly recycled in city centers:

- **Aluminum:** Rinse aluminum cans and foil that you want to recycle.
- **Glass:** You often have to sort the glass by color (green, brown, and clear). Wash jars and remove their labels. Many countries give money for certain glass containers, such as beer, wine, and soda bottles.
- **Paper:** Separate newspapers, magazines, and cardboard from regular paper and flatten boxes. If grease-stained, pizza boxes go in the garbage.
- **Plastics:** Plastics are generally categorized by numbers — including plastic bags. You can usually find these different numbers on the bottom of the containers. Check which numbers your city or town collects. Many cities and towns don't yet recycle plastic bags.
- **Tetra-packs:** You can recycle juice boxes and cartons used for milk, juice, and even wine. Just rinse and flatten them.

Throw only recyclables into the recycling bin. If your city takes only number 1 to 5 plastics, don't put in number 6 hoping it'll just get mashed in with the others. Workers at the plant see that misplaced item and often toss your whole bag in the garbage — which is more efficient than sorting through your mistake.

Where you can recycle these things depends on your area. If you have curbside recycling, you can recycle as easily as you can toss something in the garbage. Unfortunately, not all cities collect recycling yet. You may have to bring your recyclable products to recycling bins, which you can usually find at your local dump.

Composting

When you put food waste in a garbage bag, you create the perfect conditions for methane to form because when organic material decomposes in the absence of oxygen, it generates that potent greenhouse gas. If you *compost* that material — enabling it to break down into nutrient-rich material that you can use as fertilizer — you stop methane production, keep bags out of rapidly-filling landfill sites, and (best of all) create a wonderful substance that nurtures plants. What's not to love about composting?

Home gardeners have been composting their vegetable peels, plant trimmings, leaves, and grass clippings for ages. Now, many municipalities are providing compost curbside pick-up or central drop-off locations. Municipal programs typically accept a wider range of compostable materials than a home compost pile, including meat and fish products, bones, bread, pasta, paper towels and tissues, pet wastes, and disposable diapers. These additions can go a long way to making yours a zero-waste home, and the municipality can use (sometimes even sell) the compost they produce.

If you live in an apartment, or if your city doesn't pick up your compost on the curb yet, you can opt for vermicomposting. If the name makes you squirm (like it still does for Zoë), it's fitting — *vermicomposting* literally means composting with worms. A little creepy, but, with fans like Martha Stewart, it's a good thing. The little guys simply live in a bin, munch on your food scraps, and send those scraps out the other end as compost. They're quite happy to compost your scraps, and they break down your food rather quickly. You just need a bin with air holes, soil, and the worms. Or you could opt for a *worm condo,* like the one that Oprah has — a fancy stacked and aerated bin.

Chewing on Food Choices

It's time to choose a low-carb — low-carbon, that is — diet. Your food choices have a surprisingly large impact on greenhouse gas emissions. Researchers have estimated that the average American creates 2.5 metric tons of carbon dioxide emissions each year by eating, which is actually more than the 2 metric tons each U.S. citizen generates by driving.

Avoiding the big chill

The fewer cooled or frozen foods you buy, the less you contribute to the energy needed to keep all those fridges, freezers, and refrigerated trucks running. When it comes to prepared foods — ready-made meals, pizza, bread dough, pastries, and so on — simply avoid the frozen option. Frozen foods just add to the energy bill.

Sometimes, you need to chill food to prevent spoilage, reduce waste, and avoid extra trips to the store. And in the middle of winter, you may have trouble deciding whether to choose the fresh vegetable trucked from Texas or the local one that was frozen in September. The best option may be to go without your spinach or broccoli in winter, and instead get some local root crops stored without freezing.

Opting for unprocessed

Think of the extra energy it takes to turn apples into applesauce or soybeans into veggie sausages. Processed foods also need much more packaging, which uses energy and adds to solid waste streams. And the processed food you're buying probably isn't locally produced, which means that it had to be transported to you from afar, creating even more emissions.

Buying the raw ingredients for food and making it yourself cuts back on greenhouse gases. As Michael Pollan, author of *The Omnivore's Dilemma* (Penguin), recommends, "Shop at the edges of the grocery store [where they keep the natural food], not the middle."

Fresh and unprocessed foods are healthier for you, containing less salt, sugar, and mysterious chemical ingredients that no one can pronounce.

Bottled water: All wet

Bottled water is an environmental disaster. It's taken from sources hundreds and often thousands of miles away from where people will consume it, requiring enormous energy expenditures for shipping and removing the resource from its natural setting. It adds billions of bottles to the solid waste stream every year — 40 million a day, for example, in the U.S. alone. And its health benefits are questionable, to say the least: The water coming out of the tap in your municipality has been treated to meet far higher standards.

If you don't like the taste of chlorine in treated water, use a filter system. Also, if you live in an older part of your community, ask local water officials about testing your supply for lead, which was used in older supply pipes. You may need to have those pipes replaced.

Minimizing meat

If everyone in the developed world gave up meat from cud-chewing animals, such as beef and sheep, they would do more to reduce greenhouse gas emissions than giving up their cars, some research says. No wonder experts recommend eating less meat to help reduce greenhouse gases.

Even without going vegetarian, reducing intake can help. The average person needs only between 50 and 100 grams of protein a day, depending on his or her weight and activity level. You can get that from a range of foods, including fish and chicken (both easier on the environment than beef), eggs, tofu, beans, and nuts.

When choosing fish, do careful research to ensure it was harvested by using practices that support sustainability.

Here are a couple tips to help you move to a greenhouse gas–reduced diet:

- **Take a day or more each week off from meat.** Some of the world's tastiest cuisines use little or no meat. Explore vegetarian cookbooks (we recommend *Vegetarian Cooking For Dummies,* by Suzanne Havala [Wiley]) and vegetarian-cooking Web sites for ideas.
- **Choose wild or pastured meat over meat from animals in feedlots.** In feedlots, the animals are forced to consume huge quantities of grain. Raising organic beef on grass rather than feed involves 40 percent fewer greenhouse gas emissions and consumes 85 percent less energy. You can find grass-fed beef, as well as bison, in some markets (renowned climate advisor Louise Comeau makes a mean bison chili). Get game such as venison, moose, and caribou if you can find them available locally.

Buying local produce

You're probably used to finding fresh vegetables and fruits all year long, shipped from around the globe. But food that comes from close to home is so much better — it doesn't need to be transported, it helps local food producers, and it's higher in nutrients.

When you're grocery shopping, find out where the produce was grown and try to stick to nearby suppliers. Better yet, get your food directly from the source: Many cities and towns have farmers' markets, often on weekends, where you can get locally grown food that's guaranteed to knock your taste buds away. Check out great resources online for The 100 Mile Diet (`www.100milediet.org`) — all about the benefits of eating locally grown and produced foods.

Food isn't the only thing you can buy from a local source to cut back on greenhouse gas emissions. Out-of-season flowers are shipped by air or grown in greenhouses. Stick to what's in season and field-grown locally (you can find plenty at farmers' markets). During the winter, consider using dried plants and flowers, and other decorations from local sources. In spring and summer, opt for potted flowers.

Of course, eating locally is carbon-low only if you buy in season. Local food grown out of season in energy-intensive greenhouses may be worse for the environment than food that's shipped from elsewhere, recent research has found.

Choosing organic

The big O: Organic foods. Because they're grown without pesticides or artificial fertilizers, they're arguably better for the stability and health of the land than non-organic foods.

Healthy soil is especially important because earth is a major carbon sink. (Refer to Chapter 2 for more about carbon sinks.)

Organic practices are spreading, but getting certified takes time and money that can present a barrier to smaller local farmers. You may face an uncertain choice between shipped organic food and local non-organic food. Overall, a local and non-organic product will usually have less of an impact on greenhouse gas emissions than a non-local and organic one.

Wine over the waves

Not all products from afar need to arrive with a high carbon price tag. For example, bottles with a Sail Wine logo may soon be coming to a liquor store near you. For the first time in 150 years, French wines are being shipped under sail. Ireland is the first market for these wines, but at the time we wrote, the Compagnie de Transport Maritime à la Voile plans to add England, Belgium, Canada, and Sweden as destinations. Shipping on a sailboat has almost zero emissions, and the shippers claim that the wine only gets better the longer it's at sea! Eventually, this fleet of sailboats plans to transport other goods, as well, in addition to offering a few luxurious berths for merchants who want to travel with their wares.

You also may be able to go to a comprehensive, sustainable local food chain. In Canada, for example, the Local Food Plus (LFP) initiative brings together economic, social, and environmental considerations, and rewards local farmers (both conventional and organic) who use ecological practices. Local food networks exist in many places on a much more . . . well, local level. Often, communities and regions have their own local food network that can help you figure out where to buy local foods.

Cooking up fewer greenhouse gases

Energy-wasteful kitchen practices can undo even the most carbon-friendly food choices. A major food-services company has estimated that people can correct energy losses of up to 30 percent in home or commercial kitchens for very low cost.

To run a more energy-efficient kitchen, do the following:

- **Keep a lid on it!** Research has shown that the simple act of using a saucepan lid reduces energy used for simmering by a factor of five.
- **Less is more.** Use the smallest pot and least amount of water needed for what you're boiling.
- **Size your appliances to your lifestyle.** If you're on your own, you don't need a huge oven to heat a single-serving meal.
- **Use a toaster oven or a microwave.** They take a fraction of the energy of a conventional stove or oven to get the same results. And in the summer, not having to turn the stove or oven on helps keep the kitchen from heating up.

Eco-Shopping

In this chapter, we talk about better choices in electronics, appliances, and home finishing materials. But green shopping goes far beyond that — you can find low-carbon products in stores of every kind. Better choices are in the bag (a reusable cloth one, that is).

"No thanks, I don't need a bag."

Many countries and communities are moving to get rid of plastic bags, which are made from petroleum (a fossil-fuel product) and take up valuable landfill room. Grocery stores in Germany voluntarily started charging customers for plastic bags years ago. They set the price high enough (it's currently over a dollar) that consumers remember to bring a cloth one or find it more economical to buy a new cloth one if they forget theirs. Many stores in Canada, the U.S., the U.K., New Zealand, and Australia now sell and promote reusable bags. Some jurisdictions, such as Greece and Ireland, charge a tax on plastic bags, and others have imposed a ban or are planning to.

Canvas totes aren't a fashion faux pas anymore. In fact, handbag designer Anya Hindmarch created a bag emblazoned with the logo, "I'm not a plastic bag," which originally sold for five British pounds but went for hundreds of dollars on eBay after it was spotted in the hands of celebrities such as Keira Knightley.

Clothes make an environmental statement

What are the most energy-efficient clothes? The ones you already own, of course. Even though new processes and fabrics are reducing the energy needed to create clothing, every new item still comes with an energy price tag attached. So, avoid new purchases whenever possible. Treat your clothing with care and remember that patches and repairs add real character to your casual wardrobe.

Caring for your clothes

Because 75 percent of the energy consumption associated with clothing comes from laundering it, stick to cold-water washes and line-drying as much as possible. Never buy anything that needs dry-cleaning (but remember that you can wash some "dry-clean only" fabrics by hand, if you're careful). Shake clothes out before hanging them to dry and then again after to reduce the amount of ironing they need.

If you have to get rid of an item, be sure to recycle it. Old t-shirts, sweats, and flannel pajamas make great cleaning rags, and you use fewer paper towels. (Nothing beats old cloth diapers for cleaning up spills and dusting!) At the end of their useful lives, they can go into your compost pile, along with anything made of wool or hemp. Also, not-for-profit organizations in most communities collect, reuse, and resell old clothing, as well as household items.

Making smart shopping choices

If you really have to add an item to your wardrobe, you can still make choices that have less of an impact on the environment:

- **Shop for vintage clothes.** Scouring second-hand boutiques is fun and creative. Many leading-edge new designers haunt vintage shops for finds that they deconstruct and recombine into great new fashion.
- **Look for organic fabrics.** Manufacturers can grow cotton, wool, linen, and hemp organically — so look for that on the label. Raising conventional cotton alone uses 10 percent of all agricultural chemicals in the U.S. Generally, if it's organic, it's also been manufactured as energy-efficiently as possible.
- **Go for classic, as opposed to trend-of-the-moment.** No matter how it was made, a garment that lasts 20 years before it needs replacing is a better bet than one you want to toss out a year from now.

Choosing man-made fibers

Even man-made and synthetic fibers can have their place in your climate-friendly wardrobe, as long as their production and use is environmentally sustainable. Man-made fibers come from wood pulp, bamboo, soy, or corn through processes that are almost the same as those used to make polyester or nylon. The fiber equivalent of biodiesel and ethanol fuel, they're better than virgin synthetics because the raw materials come from renewable resources, and when garments made from these fibers need to be thrown out, you can compost them.

Synthetics can add durability when blended with natural fibers or provide lightweight fleece garments that allow you to more easily turn down your thermostat in winter. The longevity of synthetics can help conserve fossil fuels, lessening the need for reproduction. Also, plastic bottles are regularly recycled into fiber for clothing.

Of course, you need to return man-made fiber materials to the recycling stream when you can no longer use them so that they can feed into the process all over again. Municipal recycling lags in this area, but many innovative companies, such as Mountain Equipment Co-op in Canada and U.S.-based Patagonia, are developing programs to recycle their own and even other companies' brands. Check the Web for criteria and to look for other recyclers by using search terms such as "ethically produced," "environmentally friendly," "organic," or "recycled fabrics."

Home furnishings

If your home is energy efficient, don't you want the furnishings to match? Consider these ideas when shopping to keep your furnishings from clashing with your home, environmentally speaking:

- **Go organic.** Buy towels, sheets, and other linens made from organic fibers.
- **Consider gently used goods.** Look for stores and online outlets selling vintage table linens, quilts, and bed coverings. You have even more possibilities for reusing and recycling furniture. Older furniture is more solidly made than the modern stuff, and the energy and carbon costs associated with the wood and other materials were paid for long ago. Often, an older piece just needs reupholstering (maybe in organic linen!) to look fabulous and gain decades of renewed life. If you have a home office, look for salvaged or refurbished office furniture.
- **Look for new pieces made with old materials.** Just like in the clothing business, savvy designers are creating "new" furniture by recycling wood that would otherwise be trashed or burned.

Creating a Green Workplace

Most of the changes that you can make at home you can make at work, too. Your boss or manager may be more receptive than you think to your thoughts about energy-saving changes. Be sure you know your stuff (you're already one step ahead by picking up this book). Explain that taking these kinds of measures can be great for the company's environmental stewardship and can help save money.

Here are a few extra ideas that can help reduce the energy impact of your workplace:

- **Go for energy-efficient equipment.** Purchase the most energy-efficient models available when ordering new office equipment, such as photocopiers, printers, and computers.
- **Install a motion-sensor that activates your office's lighting.** It can help cut down on emissions, automatically shutting down when no one is left in the office.

- **Use paper conservatively.** Print double-sided and on 100-percent post-consumer recycled, Forest Stewardship Certified paper. One company that made the switch to fully recycled paper calculated that they would prevent 14.2 metric tons of carbon dioxide from being emitted into the atmosphere every year. Imagine if every company and organization did the same!
- **Get the windows in on the action.** If your workplace is one of the few remaining with windows that open, use them wisely. In the summer, leave them open overnight if you can, which can help keep the work-place cool. But when the nights are chilly, make sure someone shuts those windows before you go home. If your business is considering renovations, suggest an upgrade to more efficient windows, such as double-paned ones.

Relaxed dress code, reduced greenhouse gases

Toyota Motor, Japan's biggest company, had more in mind than happy employees when it told workers to forget about wearing jackets and ties to work for the entire summer. Office buildings were able to turn their air conditioning up a few degrees because people were dressed a little lighter. This idea has now caught on around the world. Don't be surprised if the sweater Grandma knit you becomes office dress code come December.

Part VI
The Part of Tens

"Isn't that the guy who was reporting from the glacier fields about global warming still being a theory?"

In this part . . .

In this part, we get into some fun pieces of information that you can use right away. We go over the most important things you can do to stop global warming. So that you stay excited, we offer a line-up of inspiring individuals who are playing a major role in applying solutions for climate change. (Pin-ups not included.)

Because global warming is a pretty complicated issue, many misunderstandings exist. We debunk ten of the biggest myths out there. And finally, because we know you're keen to further explore this issue, we offer ten online resources that can help keep you informed— so you can do some myth-debunking on your own!

Chapter 19

Ten Things You Can Do Today to Slow Global Warming

In This Chapter

- Making small changes in your everyday life that mean a lot
- Using your purchasing power
- Talking about climate change with others
- Working to slow climate change through your job

We don't blame you if you flipped ahead to this chapter before reading all, or any, of the rest of the book. In fact, we congratulate you — you want to do something about climate change. What you can do to help slow global warming depends on where you live, the resources you have, and how much time you can give. You may not be able to slap solar panels on your roof tomorrow, and you likely can't ditch your car for a hybrid by next Tuesday. But you can make simple changes that have a big impact. This chapter offers some solutions that you can implement right away. For even more ideas, check out Chapters 17 and 18.

Driving Smart

Fuel emissions from transportation account for about 16 percent of global greenhouse gas emissions (or 24 percent of emissions from energy use, not including deforestation). Transportation also accounts for almost two-thirds of oil use in the U.S. As painful as it sounds, you can best cut down your car emissions by not owning a car at all (although without significant changes in land-use planning and access to mass transportation, this will be more difficult for some than others!).

If you must drive, then drive smart. Efficient driving is climate-friendly driving. A lot of little changes can substantially cut your car's emissions:

- Carpool when you can.
- Turn off your car when you stop for ten or more seconds. This applies when you're pulled over to stop, or delayed by traffic construction that brings everyone to a standstill — not if you are in the middle of traffic at a stoplight.
- Do all your errands at the same time, instead of spreading them throughout the day.
- Keep your car's engine clean, up-to-date, and running efficiently by taking your car in for regular checkups (or doing it yourself).
- Fill your tires to their ideal pressure point (which you can find written right on the tire) to use less fuel.
- If you have a ski rack mounted on the roof of your car, remove the rack in the summer to reduce drag, which helps reduce fuel use.

For more tips on reducing your auto emissions, see Chapter 17.

Supporting Clean, Renewable Energy

You can support the development of clean, renewable energy in a number of ways, depending on where you live.

Here are the two most common ways to make a significant impact:

- **Make the energy.** You can make energy yourself by using options such as solar energy to produce hot water and generate electricity. Chapter 13 covers other options that you can use to make energy.
- **Buy the energy.** You can purchase energy from a company that uses low-emission energy sources. Green power can be purchased in most provinces in Canada. Companies also cover 30 states in the U.S., with similar companies across the U.K., and in every state in Australia (see Chapter 18 for Web site resources on green energy providers).

Buttoning Up Your House

The key to creating an energy-efficient house (whether you want to keep the heat in or out) is insulation, insulation, insulation. Heating and cooling costs make up a whopping 80 percent of your energy bill, so seal up your doorways and windows, and make sure that you have the proper insulation for your crawl space, attic, and walls. Some of these changes might take a bit of time and effort, but they last for decades and save you significantly in energy costs, all while cutting back on greenhouse gas emissions.

Start with an energy audit so that you know where your dollars can give you the best results. Chapter 18 shows you how to perform an energy audit.

Bringing Climate Change to Work

The changes you make at home are important, no question. But most folks spend the majority of their waking hours at work. Your employer's building and transportation fleet (if your company has one) presents two major projects that are just waiting to be tackled. If you nose around, you may find that your colleagues also want to turn your office green.

Here are a few suggestions that you can make to your boss, building manager, or other decision-makers at work to get started:

- Establish an energy policy that cuts down on the company's or organization's energy use and sets a target for reducing emissions. See `www.theclimategroup.com` for a case study that might relate to your workplace.
- Put a recycling and composting system in place.
- Post signs or notices that remind people to turn off their computers, lights, and other office equipment when they leave.
- Organize monthly lunch programs and have a speaker come in to talk about solutions to climate change.

Many of these tasks mean taking an issue to a project manager or a boss, or even to the director of maintenance or facilities management. Forming a committee or advisory team with your colleagues can make you more effective and successful — you can find strength and support in numbers.

When a company stands for more than profits, employees show greater corporate loyalty and improved morale, which results in better job performance and a drop in absenteeism.

Going Vegetarian or Vegan (Sort Of)

Eating fewer meat products can really help reduce your carbon footprint. Over the course of one year, the impact of going vegetarian is similar to trading in your regular car for a hybrid car. Taking that a step further, going vegan is like trading in your regular car for . . . well, no car at all.

You don't need to go full-out vegetarian. Try starting with going one or two days per week without meat. The next best solution is to try to buy locally raised or wild meat, which usually comes from smaller, less energy-intensive operations. Chapter 18 covers more options for reducing your carbon footprint through your food choices.

Buying Energy-Efficient Electronics and Appliances

Be conservative with your purchasing choices. When you buy electronics and appliances, don't buy what you don't need. If you get a bunch of gadgets that you're lured into purchasing, you end up consuming more electricity than you intend.

But you can't cut every corner, and sometimes you just need a new appliance or electronic. The good news is that almost every kind of electronic and appliance, from dishwashers to computer monitors, comes in energy-efficient models. Start the habit of looking for energy-efficiency labels, such as Energy Star. Chapter 18 shows you how to sniff out the most energy-efficient products in your country.

The energy habits that you have in your daily life make a difference, as well. Here are a few energy-saving changes that you can make to your daily routine:

- Use a drying rack or a clothesline, rather than the clothes dryer. If you do use the dryer, fill it up and don't run it longer than necessary.
- Switch to compact fluorescent light bulbs or light-emitting diode (LED) bulbs, and turn on the lights only when you need them.

- Shut down your computer when you're not using it.
- Buy products that are recyclable or that include recycled content.
- Choose energy-efficient forms of travel, such as the train or bus, for short-haul trips. To get to work, take your bike, the bus, or the subway. Avoid unnecessary trips altogether. (For more about reducing your transportation-related carbon dioxide emissions, check out Chapter 17.)

Any choice you make, whether you're at the checkout counter or adjusting your thermostat at home, can help reduce your emissions.

Launching a Local Campaign

By launching a local campaign, you can help raise awareness and work with others to create positive change in your community. You don't have to organize a protest or stage a sit-in; you simply have to come up with a plan that's tailored to your community's and campaign's needs.

Here are some steps you can follow to help you get started:

1. **Choose an issue relevant to your community that's linked directly to climate change.**

   For example, perhaps your town doesn't have anti-idling bylaws, and you think it should. By taking on this issue, you can help fight climate change and protect kids who have asthma.

2. **Contact local organizations that are connected to or working on your issue, and ask them for resources and advice.**

   Toss a wide net — check out community service groups, youth groups, environmental groups, and so on.

3. **Find a meeting space.**

   Look for somewhere free or inexpensive, such as a public library or a community hall.

4. **Select a date for your meeting.**

   Choose a time that you think works with most people's schedules. Also, you may want to offer child care. And think about accessibility for the physically challenged.

   Give people advanced notice so that they can make arrangements to attend the meeting.

5. **Reach out to the community to get as many people as possible to attend the meeting.**

   Send event information to radio stations and newspapers. Have friends put up signs and send out invitations. Send mass e-mails. Set up a group on a social networking site such as Facebook. Don't be shy about getting the word out.

6. **Speak your piece at the meeting, and then ask everyone else to talk.**

   Decide on your concrete goal, determine an action plan, and make sure that everyone has something to do when they leave.

Good luck, and let us know how it goes!

Writing to Your Leaders

Believe it or not, letters to your elected representatives make a difference. Postcards might catch someone's attention, and petitions sometimes get noticed, but letters are by far the most effective. Politicians are eager to know what the people think.

Don't worry about composing a long or deeply profound letter. Use the letter to express your concerns and ideas as simply as possible, then send it in. To get more bang for your letter-writing buck, copy the letter and mail it to other local, regional, and national politicians. You can even request to meet with a politician to discuss what the government is doing about climate change.

You can expand your outreach beyond political leaders. Writing to people within your industry, your business, or your children's school, or even writing to your own boss, can lead to profound change.

Spreading the Word

Whether you love to talk (which we do!) or hate to, you can spread the word. Formal presentations can effectively get the message on climate change across to your family, friends, co-workers, and community.

You can either give a presentation yourself or ask someone to come in and give it for you. Al Gore's campaign and non-governmental organization called The Climate Project can show you how to present the same slide-show presentation that Gore gives in *An Inconvenient Truth.* So far, they've trained 1,000 presenters in the United States, 250 in Canada, and another 400 across the United Kingdom and Australia. The Project's always planning more training sessions, so check their Web site (`www.theclimateproject.org`) for updates. From the site, you can request a presentation for your community or group.

You can also contact any local organization that's working on climate change to request a presentation. Who knows — maybe, after you get some pointers from the local organization, you can give the presentation yourself.

Getting (Or Making) a Green Collar Job

Many jobs offer the opportunity to contribute directly to climate change solutions. And you're not limited to working within an environmental non-governmental organization (NGO) — though we highly recommend this option! The following list gives you a glance at the kind of climate-friendly jobs that are out there today, for a range of qualifications, skills, and interests:

- **Architecture and design:** The world needs green building construction and design, and this field is looking for architects and designers with vision. You can help make everything within a building — from materials to energy systems — green. (We talk about greener buildings in Chapter 14.)
- **Building/contracting:** You can offer your clients Forest Stewardship Council certified wood (see Chapter 14), materials (such as plastic, stone, wood, insulation) made with recycled content, and environmentally sound materials (that are biodegradable or able to be recycled at the end of their lifecycle) to help them make informed choices and help reduce emissions adding to climate change.
- **Education:** As a teacher at any level, you have many opportunities to integrate climate change issues and solutions into your curriculum. You can even join one of the many organizations of educators who work together to have environmental issues put into standardized public education.
- **Engineering:** Engineers who want to work in climate-related fields are in high demand. The long list of fields includes environmental assessments, water resource management, carbon capture and storage, and greenhouse gas assessments.
- **Government:** With more government programs starting up in various countries, new positions are created regularly. Look for posts within the departments that deal with climate change, the environment, energy, sustainable development, agriculture, and forestry. (Chapter 10 outlines some of the actions governments can take to help fight global warming.)

- **Higher education:** Being a student isn't exactly a job (although you can find many climate-related research grants available), but it presents a way to take your knowledge about climate change to a higher level. Many universities and colleges around the world now offer programs that focus on energy, the environment, climate change, environmental management, climate law, and more.
- **Music:** When you're a musician, you have people listening to you all the time (or, at least, that's your goal). You can heighten climate change awareness through your lyrics or by speaking to your audience between songs.
- **Non-governmental organizations (NGOs):** The campaigns and projects available through NGOs allow you to work directly on an issue and advocate for change. Whether you lead the campaign or assist in the office, your job contributes to the overall effort of effective projects. (Check out Chapter 15 for more about NGOs.)
- **Visual art:** Being a visual artist gives you a unique opportunity to express anything and everything about climate change. Photos, film, and all kinds of visual works of art can effectively communicate the urgency of global warming.
- **Writing/journalism:** If you write professionally, people read what you have to say. You have a wonderful opportunity to introduce your readers to climate change issues. (Chapter 16 explores how the media covers global warming.)

You can find endless opportunities, whether you're an entrepreneur or a job seeker, if you go looking for a green collar job. Making a difference is a terrific feeling.

Chapter 20

Ten Inspiring Leaders in the Fight Against Global Warming

In This Chapter

- Leading the way in government
- Writing the climate change story
- Demonstrating the difference activists can make
- Looking at the scientists, not just the science
- Leading the corporate fight against global warming

Thousands of people around the world are working on climate change issues and making a difference. The ten we profile in this chapter are the cream of the crop. But, hey, remember — there's always room for one more!

The Politicians

Around the world, politicians are struggling to decide how to address global warming. Despite facing budgetary challenges, resistance from industry, and a public concerned with countless other pressing issues, some leaders have managed to keep the climate crisis atop their agendas and are making positive changes. At the top of the heap are Angela Merkel and Arnold Schwarzenegger.

Angela Merkel

The first female Chancellor (or Chancelloress, as the German federal government officially notes) of Germany and the leader of the world's third-largest economy, Merkel played a key role in putting climate change at the top of the agenda at the 2007 G8 meetings, where she acted as president. She also held the position of president of the European Union in 2007 and used that position to push for the adoption of the European Union target of a 20-percent increase in energy efficiency by 2020. Merkel served as Germany's minister of the environment in the 1990s, during which time she played a key role in convincing developed countries to join the Berlin Mandate of 1995, which led to the creation of the Kyoto Protocol in 1997. Kyoto may not have happened without her work.

Speaking on climate change to the United Nations General Assembly in 2007, Merkel said, "It's not only the dry facts and numbers that call us to negotiate, but also the question of in what future we want to live. For me, this is not only an economic imperative, but also a moral one."

Arnold Schwarzenegger

Known worldwide as an action hero in such movies as *The Terminator, Commando,* and *Predator,* Arnold Schwarzenegger took on a new role in 2003: governor of California. He has since been at the forefront of sweeping political change, making the state of California a leader on climate change. He introduced green, climate-friendly innovations such as the Hydrogen Highway (putting the infrastructure in place to allow people to use hydrogen fuel-cell cars across the state) and the Million Solar Roofs Plan (a statewide solar rebate program helping to make possible 1 million solar roof installations). You can see his dedication to dealing with climate change in legislation that ensures cost-effective greenhouse gas reductions by using principles of business and economic policy. (Check out Chapter 10 to read about some of California's success stories.)

The Wordsmiths

When it comes to fighting climate change, the pen is definitely mightier than the sword. The works of the two writers discussed in the following sections have topped bestseller lists worldwide and affected everyone who's read them.

Tim Flannery

An internationally recognized scientist, writer, and explorer, Tim Flannery's most recent book, *The Weather Makers: How Man Is Changing the Climate and What It Means for Life on Earth* (Atlantic Monthly Press), received international acclaim for its in-depth look at climate change. His skill at clearly communicating the importance of the climate change problem has increased the worldwide feeling of urgency to address climate change. *The Weather Makers* was the first book to address the major questions on so many people's minds: How serious is climate change? How much do we know? What's going to happen? What can we do?

The book publicized the full story of climate change around the world by making the issues of climate change accessible to everyone. His dedicated work earned him the Australian of the Year award in 2007.

George Monbiot

U.K.-based writer George Monbiot is best known for his columns in *The Guardian* newspaper and his recent best-selling book, *Heat: How to Stop the Planet from Burning* (South End Press). George Monbiot's clear and compelling writings have brought attention to the perils of climate change. Monbiot has been praised for his forthright and demanding ideas for greenhouse gas reductions, developing what appears to be a feasible plan for a 90-percent reduction in greenhouse gas emissions by the year 2030. Monbiot demands the attention of anyone interested in climate change.

The Activists

People worldwide recognize global warming as a major issue thanks to the efforts of activists around the globe who are constantly reminding the media, the government, and the public about the climate crisis. The four individuals in the following sections, from very different backgrounds, demonstrate the many ways we can all fight global warming, and how just one person can truly make a difference.

Al Gore

Former U.S. Vice President Al Gore has been an advocate for environmental issues for most, if not all, of his life, and he has actually done more for the fight against climate change as an activist than as a politician. He, along with the UN Intergovernmental Panel on Climate Change (IPCC) scientists, was awarded the Nobel Peace Prize in 2007 for that activism.

Today, Gore is probably best known for the documentary (and following book) *An Inconvenient Truth,* based on the presentation he gives worldwide on the urgency of climate change. This film led to the creation of The Climate Project, a group that trains individuals around the world to give the acclaimed presentation seen in the documentary. Gore also organized the international Live Earth benefit concerts, which raised the climate change awareness of the more than 1 billion people who tuned in.

He has built on this momentum by spearheading the We Can Solve It campaign, aimed at further educating, engaging, and mobilizing people around the world to take immediate action on climate change. This campaign is a project of The Alliance for Climate Protection, a non-profit and non-partisan organization. Gore continues to tell the world that climate change is an urgent problem, requiring an urgent and inspired response.

Wangari Maathai

Wangari Maathai is best known for founding and building the Green Belt Movement, a group whose main focus is to plant trees to restore the natural environment. Starting in Kenya, the movement has since become international.

A pragmatic and effective organizer, Maathai was often on the wrong side of the Moi government, serving time in jail and under threat for her work. Her amazing life of achievements includes being the first woman in East and Central Africa to complete her doctorate, receiving the Nobel Peace Prize, and serving on the Kenyan Cabinet. Most importantly, she has effectively mobilized millions of individuals to be a part of the solution. At the top of her current to-do list is to plant a billion trees in Africa, helping African countries adapt to the effects of climate change.

Sheila Watt-Cloutier

A long-time advocate for Inuit rights, Sheila Watt-Cloutier speaks out today on climate change. She and her community, in fact, are already feeling its effects. Originally an advocate against contaminants in Arctic wildlife that the Inuit depend on, her focus expanded to climate change advocacy when she became the chair of the international Inuit Circumpolar Conference. Watt-Cloutier has made the world understand that climate change is inextricably linked with the survival of Inuit culture and spirit. In 2007, Watt-Cloutier was nominated for the Nobel Peace Prize and garnered the Canadian Lifetime Achievement Award for this work.

While she was president of the Inuit Circumpolar Conference, Watt-Cloutier brought a lawsuit — that continues to this day — against the United States to the Inter-American Commission ("Court") on Human Rights, alleging that the U.S. government's rejection of the Kyoto Protocol is threatening the Inuit with cultural genocide.

The Scientists

The scientists got the ball rolling with regard to climate change awareness, and their research continues to lay the groundwork for the world's understanding of climate change. Without the dedicated work of researchers, the world wouldn't know about the effects that our actions have on the planet.

James Hansen

People around the world, who value courage in telling the truth, despite extraordinary pressure and threats, applauded James Hansen, scientist and advocate for urgent solutions to climate change, for speaking up when the U.S. federal government tried to censor his work with NASA on climate change. This censorship is the subject of Mark Bowen's book *Censoring Science: Inside the Political Attack on Dr. James Hansen and the Truth of Global Warming* (Dutton).

Although Hansen works as a physicist on major climate change modeling, his work outside the office has had the greatest impact on the fight against global warming. He has made it a personal mission to communicate climate change to the public in the most clear-cut way possible.

Rajendra Pachauri

Pachauri is the current chair of the IPCC, the volunteer-based group that shared the 2007 Nobel Peace Prize with Al Gore for work on climate change. The pressure has been on Pachauri since 2002 because his position as chair has never been more important.

Pachauri, along with many other leading IPCC scientists, attends all the UN Climate Change Conferences and delivers presentations explaining the latest science to negotiating bureaucrats. He also writes IPCC reports about the science of climate change for politicians and the public.

The Business Leader

As one of the crusaders in environmental business leadership, Ray Anderson has brought his role as founder and former CEO of Interface Flooring to a new level. His vision, and now his company's commitment, is to eliminate *any* impact it has on the environment. And it's already halfway there. As he puts it, he's climbing Mount Sustainability.

Starting from a midlife, late-career revelation after reading Paul Hawken's *The Ecology of Commerce* (Collins), Anderson set out to leave the world a better place — not just leave a lot of money and a big corporation. He shifted his company's annual statement to "what we take, what we make, and what we waste" accounts.

His central discovery was that every time Interface closed off a smokestack or a wastepipe, the company saved money. Every energy-saving effort, including tapping into landfills for the energy to run Interface's plant, improved the corporate bottom line.

In his personal life, Anderson drives a hybrid car and lives in an *off-the-grid* home (the house doesn't draw energy from local power supplies but generates its own renewable energy). Professionally, he continues to share his best practices with businesses around the world that are looking to become more sustainable and lower their impact on climate change.

Chapter 21

Top Ten Myths about Global Warming

In This Chapter

- Knowing that a debate doesn't exist among scientists
- Recognizing that global warming doesn't result exclusively from natural climate variation
- Looking into the danger of increased carbon dioxide concentrations
- Discovering the truth about sunspots
- Understanding that scientists don't exaggerate to get more funding
- Grasping the misconceptions about the science of global warming
- Pinning the blame on developing countries isn't realistic
- Debunking the benefits of the Northwest Passage
- Living with climate change — and doing something about it
- Predicting the future by using scientific models

Global warming has been a topic of discussion for years, but the discussion only recently has permeated all countries and cultures. In some cases, the fossil-fuel industry has financed major public relations campaigns to promote doubt about the level of risk and degree of scientific consensus concerning global warming. Because of their efforts, a great deal of misunderstanding exists. In fact, if you ask some folks, they'll tell you that the whole "climate change thing" is nonsense. In this chapter, we tackle the ten most common myths — and offer arguments that refute them.

A Big Scientific Debate Exists

The idea that a big scientific disagreement exists regarding global warming is one of the most persistent and erroneous claims used to delay taking action. For more than a decade, the overwhelming majority of scientists have agreed on the key elements of the problem:

- The planet is and has been warming, and it will continue to warm for the foreseeable future.
- The warming isn't happening because of natural factors alone; it's largely due to human activity — burning oil, coal, and gas, and destroying forests.
- The impacts of the climate changes that are happening because of rising temperature are serious and represent a significant threat, requiring global action.

The media spotlights many scientists who claim that doubt exists within the scientific community regarding global warming. Very few scientists foster that appearance of debate, and those who do are by and large not active in the field of climate research. Cigarette companies used similar tactics, finding doctors to dispute that smoking caused lung cancer. Al Gore even uses an old cigarette ad that shows doctors recommending a particular brand of cigarette in the documentary *An Inconvenient Truth.*

The Intergovernmental Panel on Climate Change (IPCC) is the most credible source of climate-related information; their reports reflect consensus among the scientific community. If anything, the IPCC underestimates the severity of the risk of the coming climate crisis.

The Warming Weather Is Natural

You may hear global warming skeptics say that weather goes in cycles and the recent warming trend is just that: a trend that will correct itself. Natural climate variation does exist — that much is true. Scientists have records of wide temperature variations over time, ranging from the temperature lows of the Ice Ages to many millions of years ago when the Arctic was a swampland. A combination of temperature, greenhouse gas emissions, plate tectonics, and the sun affect the climate.

But all those things considered, scientists have reached a consensus that natural cycles can explain only a part of the current warming. Today's quickly rising temperatures are unlike any recorded changes. Warmth isn't inherently a bad thing, but because the shifts in temperature and regional climates are happening so suddenly, plants and animals probably won't be able to adapt.

Some people use what seems like a commonsense approach to explain away global warming. They point out that humans didn't keep good temperature records thousands or even hundreds of years ago. True. But the data extends further back than the approximately 140 years that people have been keeping temperature records. Scientists can make reliable estimates about the baseline temperature trends by using indirect measurements from sources such as ice cores from glaciers and tree rings from ancient forests.

Science that bases its conclusions on concrete evidence — from ice core samples to melting glaciers — makes a clear case that the warming of the modern world is well beyond natural variation. The atmospheric concentrations of the greenhouse gas carbon dioxide are now about 35 percent above the highest levels in the last 800,000 years (based on direct measurements from ice cores), and have likely never been this high for over 20 million years.

Carbon Dioxide Isn't a Big Factor

Carbon and temperature are not always linked in their historical records. But it doesn't make sense to take a bit of the truth, such as this fact, and use it to create the impression that no link has been proved between human-caused emissions of carbon dioxide and changes to global temperature (and hence the climate). Sure, over a geologically long period of time, carbon dioxide levels and temperature aren't always in lockstep. Carbon dioxide is only one of many greenhouse gases and only one part of the climate equation. However, carbon dioxide is the greenhouse gas that's playing a lead role now.

Carbon dioxide is a powerful warming gas. About 280 parts per million of carbon dioxide kept the planet livable for millennia. That many parts per million is like a drop in a swimming pool. When the concentration of carbon dioxide increases, the atmosphere heats up significantly. That said, an increase in carbon dioxide isn't the *only* thing that's ever heated up the atmosphere.

Carbon dioxide is just one of many factors that influence the world's temperature. Overall, however, the more carbon dioxide you emit into the atmosphere, the warmer the Earth gets.

Global Warming Is Caused by Changes in the Sun

A number of factors influence the temperature on the planet, including *solar flaring* (or sun spots) and *radiative forcing* (the impact of the sun's activity, which varies, on the climate). The IPCC scientists reviewed the literature about the possibility of sunspots and changes in the sun's activity affecting the planet's climate. They estimate that the warming effect from increases of greenhouse gases in the atmosphere is more than eight times greater than the effect of solar irradiance changes.

Scientists Exaggerate to Get More Funding

Cynics charge that scientists overstate the urgency of global warming in order to obtain more grants. The notion that the world's leading scientists and scientific academies fudge the evidence and put themselves at odds with the most powerful corporations on Earth for money is pretty far-fetched. If anything, the environmental community and scientists have been too cautious in expressing concerns.

The anticipated changes documented in the early IPCC reports, starting in 1990, were all based on what would happen if the carbon dioxide emissions doubled, which led some to think that the only risk was if the concentration of greenhouse gases doubled to 550 parts per million (ppm). Now, scientists increasingly understand that the risk to societies and ecosystems is unacceptably high at levels far below 550 ppm. Increasingly, scientists are calling for stabilization of carbon dioxide emission below 425 to 450 ppm — the atmosphere is at 387 ppm already, and rising at approximately 2 ppm per year. They say that industrialized countries need to cut greenhouse gas emissions by 30 percent by 2020, leading to 80-percent reductions by 2050. Globally (including developing countries), humanity needs to cut greenhouse gas emissions by 50 percent below 1990 levels by 2050.

Science and Technology Will Fix It

You may hear people talking about new technologies that can save the world — clean coal, a hydrogen economy, or carbon capture and storage. Stick around long enough, and you even hear really wild sci-fi solutions, such as adding iron oxide to the ocean or putting mirrors in outer space. Studies show that many of these technologies, especially forms of *planetary engineering* — in other words, large-scale re-engineering of the planet — would make things worse.

Still, high-tech solutions are available, and you're bound to have seen new breakthroughs well before this book made its way to your local bookstore. Each potential solution, especially the proposed use of hydrogen and fuel cells, may play an important role in the future, but none of them justify waiting and postponing the actions people can take right now.

Solutions that sound too good to be true are probably just that — too good to be true. No solution allows humanity to have its cake and eat it, too. People need to reduce emissions fast. Really, everyone needed to reduce those emissions yesterday.

Developing Countries Will Only Make Matters Worse

It doesn't matter what developed nations do, critics of the Kyoto Protocol say, because the developing nations are totally unregulated. Look at China — they're already the world's biggest polluter.

To make an accurate assessment, you must consider greenhouse gas emissions per person. Sure, China is the now a slightly higher emitter than the U.S., but at 4 metric tons of carbon dioxide emissions per person, it produces less per person than 72 other countries. India is even lower at 1.1 tons per person, says the World Resources Institute. In 2004, Australia and Canada, on the other hand, produced 17.9 and 17.3 metric tons per person. The United States was even worse at 20.5 tons per person.

"Rich guys go first" is the name of the game. Industrialized countries have the resources available to reduce emissions and are responsible for the majority of emissions. Requiring developing countries to reduce emissions without the support of industrialized countries would saddle those countries with a financial burden they can't bear and without the tools to meet the challenge. Industrialized countries must lead the way on emission reductions so that developing countries don't follow the same development path that the industrialized countries took — one based on fossil fuels.

Vanishing Arctic Ice Will Help the Economy

Many people see at least two benefits of the vanishing Arctic ice. First, the legendary Northwest Passage could be seasonally open — when the ice melts — any summer now. Folks in shipping like the look of this shortcut between the Atlantic and Pacific oceans. The Northwest Passage can cut thousands of miles off the route now followed through the Panama Canal. The loss of ice also invites the idea that fossil-fuel companies can tap oil and gas reserves lying under the Arctic Ocean. Yes, companies want to extract and burn the very fossil fuels whose emissions caused the ice to melt — causing more ice to melt.

Human-induced climate change has already caused the loss of 1.2 million square miles (2 million sq km) of ice — a 10 to 20 percent decline of summer sea ice cover, as compared to what was present 30 years ago. The remaining ice is also thinner than it was 30 years ago. These trends of thinning and melting accelerate the rate of warming in what scientists describe as a *positive feedback loop:* While the ice melts, the *albedo effect* (ice reflecting the sun's radiation away from the surface) is reduced, and the open dark waters absorb the sun's heat, leading to faster ocean warming and more melting ice.

The resulting changes in ocean temperatures have global impacts. Without question, melting Arctic ice threatens the world as we know it. Surprisingly, rising sea levels aren't the threat. Rising sea levels occur when the ice of Antarctica and Greenland, where ice doesn't lie over open ocean, melts and runs into the ocean.

The Arctic ice melting is like the ice cube melting in your drink on a hot day. The melting ice takes almost the same space as the frozen ice. Because Arctic ice is made of fresh water, it stays above the saline waters of the ocean when it melts, depressing the current. The Gulf Stream has already slowed by 30 percent in the last few decades.

While more Arctic ice melts, the Gulf Stream is at greater and greater risk. If the Gulf Stream stalls, the world will experience a massive and abrupt climate shift. The resulting changes would move like a row of dominos, causing sudden changes in precipitation patterns, drought in major food-growing regions, and shifts that increase severe extreme weather events. The U.K., for example, would find itself in a much colder climate, more like that of Canada.

People Can Adapt

That humanity can adapt isn't really a myth. People *can* adapt. But humanity's adaptability doesn't mean people can ignore the demands to reduce greenhouse gases.

People can adapt to the impacts of climate change at the lower end of impacts. In other words, scientists know that the climate is changing because of human activity. But humanity doesn't know where the new concentrations of greenhouse gases will level off. What will be the new normal?

If the world can reduce greenhouse gases quickly enough, humanity could hold that new normal to 400 to 425 ppm. Many scientists are arguing that we must keep carbon at no more than 350 ppm. But if people adopt the "Who, me? Worry?" philosophy, the new greenhouse gas concentrations could level off at 550 ppm, 650 ppm, 700 ppm, or higher.

By 450 ppm, most scientists anticipate levels of climate crisis so severe that humanity won't be able to adapt. These predictions paint a picture of a very grim future. Humanity would have to deal with natural disasters, rising sea levels, and social and political disruption; millions of environmental refugees; and food shortages and famines. No one can adapt to that.

So, humanity needs to reduce emissions, but people also *must* adapt.

Scientific Models Don't Accurately Project the Future

No one can perform traditional science on the entire atmosphere. You can't, for example, manipulate one variable, observe how all the others change, and then jump in a time machine and re-run the experiment a few times. Instead, scientists use computer simulations that rely on well-established scientific equations that describe the atmosphere and oceans.

Some critics claim that people don't and can't ever fully understand the complexity of the climate. Of course humanity won't *fully* understand. As it turns out, however, the present understanding is more than adequate to act.

Some folks expect scientists to invent a functional crystal ball. Of course, no crystal ball can tell your fortune, but it's amazing how close scientists have come to projecting the future. The models they use are incredibly intricate and require super-computers to crunch through the data. Those computers give highly effective models.

Chapter 22

Ten Online Global Warming Resources

In This Chapter

- Helping your kids understand global warming
- Taking a deeper look at the science
- Looking at your country's role in the solutions
- Exploring energy issues
- Going global

Although global warming has just recently caught the world's attention, you can already find a lot of information about it online. But you might find sorting through it all challenging. In this chapter, we point you to the best sites, whether you want to explain to your kids about global warming, need a place to start for your own research, or want to know as much as you can about the issue.

Never Too Young: EcoKids

You can help the kids in your life understand the environment and climate change by pointing your computer's Web browser to the EcoKids site (`www.ecokidsonline.com`). "Teach a child, and you teach the world" is the site's slogan. Although EcoKids is based in Canada, everyone can find it useful, regardless of where you live. The site boasts more than enough content to keep any kid happy and discovering interesting facts about climate change for hours. EcoKids includes the following content:

- **Games:** Each game shows kids how to make the best choices about climate change and the environment.

- **Homework help:** EcoKids offers interactive information sites to help kids with homework related to climate change, energy, land use, and water, among other topics.
- **Content for teachers:** With resources for in-class curriculum, book lists, and resources for starting an EcoKids club, teachers find the site a dream.

Sticking with the Science

The amount of scientific research on global warming is constantly growing, and scientists are always updating their findings. To get in-depth information and the latest research on an ongoing basis, we recommend the following Web sites:

- **Intergovernmental Panel on Climate Change** (`www.ipcc.ch`): Created in 1988 by the World Meteorological Organization and the United Nations Environment Program (UNEP), the Intergovernmental Panel on Climate Change (IPCC) is a team of scientific experts who, among other duties, writes reports for the United Nations. These experts look at all the peer-reviewed science in the world relating to global climate change and synthesize it into assessment reports. On their site, you can access their numerous reports, which cover the science behind climate change, its effects, and potential solutions to it.

If you want the detailed scientific background to climate change, check out the IPCC's full reports; if you're just looking for the big picture, we recommend the reports for policymakers.

- **Climate Analysis Indicators Tool** (`http://cait.wri.org`): This site, offered by the World Resources Institute, enables you to create graphs that compare greenhouse gas emissions by country, by source of emissions, by carbon dioxide per person by country — or by pretty much any other combination you can think of, actually. You need to sign up in order to use the tool, but it's free.

  You may find the site a little intimidating at first glance because it's not geared to the general public. Just dive in and use the tool, however, to become more comfortable.

Going Governmental

Your own government's official Web site often has current information on climate change projects and initiatives underway, how this information affects your life, and what you can do to become involved. But don't feel bound by borders — check them all out!

Australia

The Australian government has an official Greenhouse Office. (Bet they get to wear shorts to work all year round.) On their site (www.greenhouse.gov.au), you can explore

- The connections between climate change and Australia's natural resources, industry, and agriculture
- Global warming's impacts on Australia and how the government plans to deal with the problem
- How to shop for or build a climate-friendly home

This Web site is the most comprehensive of the bunch listed in this and the following sections, and it's a great resource for anyone. Nothing is missing! (Not even, since the recent change of government, a Kyoto Protocol target.)

Canada

Environment Canada's simple climate change Web site, www.ec.gc.ca/climate, offers a brief overview of climate change, suggests actions for you to take at home, and provides access to speeches about climate change made by government officials. By digging around the site a bit, you can also find the following information:

- Reports on Canada's greenhouse gas emissions and commitments to the Kyoto Protocol
- How to apply for incentives and rebates that apply to activities such as renovating your home, upgrading your car, or doing pretty much anything else that improves your personal energy efficiency
- Examinations of climate change and its relation to agriculture, energy, and transportation in Canada

The Canadian government also hosts a site dedicated to greenhouse gases at www.ec.gc.ca/pdb/ghg. For a more general environmental site run by the government, visit www.ecoaction.gc.ca.

United Kingdom

The British government's climate change site (www.climatechallenge.gov.uk) is our personal favorite because it's clear, direct, and to the point. At this comprehensive Web site, you can

- Bone up on the basics of global warming
- Calculate your own carbon dioxide emissions
- Look into the climate change projects, initiatives, and policies of the federal government
- Get involved in local projects

United States

The U.S. Environmental Protection Agency's site offers a bunch of useful information. By browsing around `www.epa.gov/climatechange`, you can find the following:

- The government's climate policy
- The effects of climate change on health and the environment
- Links to your state and local governments, as well as their action plans

We've Got the Power: International Energy Agency

If you get energized thinking about the possibilities for power that exist beyond fossil fuel, check out the Web site for the International Energy Agency (IEA), `www.iea.org`. The IEA acts as an energy consultant for its 27 member countries, which include the United States, Canada, Australia, and the United Kingdom. Originally created during the energy crisis of the mid-1970s to deal with oil supply emergencies, the IEA's focus has changed with the times, and it now deals with all sources of energy.

The IEA's Web site is for the ambitious. If you want to know the ins and outs of the global energy world, this is the place to find it. Throughout the entire site, you can find information on the following:

- **All forms of energy:** The IEA offers a wealth of reports about energy sources such as renewable sources, natural gas, and oil. Select the energy of your choice from a list of links under the Topics tab, and the site displays information on publications, programs, workshops, and even contact information if you have questions.
- **Specific energy issues:** The IEA provides information about how energy is relevant to many different subject areas, such as sustainable development, emissions trading, and the Clean Development Mechanism of the Kyoto Protocol.

- **Country-by-country data:** The IEA Web site indicates how much energy each country uses, where this energy goes, and the source it comes from. You can select a country name from a list or click a region on the site's map. Each country's page contains links to the amount of oil and type of renewable energy used by that country, as well as the country's energy policies.
- **How the IEA works:** The site explains how the IEA works with energy producers, industrialized countries with high energy demands, and developing countries (such as India and China) to try to make energy supplies available and sustainable in the face of climate change.

The IEA also operates a separate site about carbon capture and storage, a topic that we cover in Chapter 13. You can get into the nitty gritty on that topic at `www.co2captureandstorage.info`.

Thinking Globally

Although climate change is truly a global problem (which is probably why it's also called *global warming*), much of this book focuses on the United States, the United Kingdom, Canada, and Australia. If you're looking for a bigger picture, we recommend a visit to the sites in the following sections.

Gateway to the UN System's Work on Climate Change

The United Nations is a giant organization, with an array of sub-organizations, many of which are working on global warming issues. Point your browser to `www.un.org/climatechange` to explore the goings-on of these organizations.

The organizations that appear on the left menu bar are UN departments that have climate change projects — it's not a short list! Click one of these links to open the climate change Web site of that department. These UN sites demonstrate the connections between climate and other important international issues, such as food, human settlements, and economics.

Beyond the links to these programs and organizations, this site offers the following:

- **International climate change projects:** A list of incredible climate change projects, categorized by country and led by various UN departments

- **Resources for children and youth:** Games, programs, and event listings relevant to young people
- **General climate change information:** Fact sheets that contain information regarding climate science, politics, and actions to take

The Pew Center on Global Climate Change

Although based in Arlington, Virginia, the Pew Center on Global Climate Change (`www.pewclimate.org`) has a truly international perspective, as its name makes clear. The center produces highly acclaimed research on almost every topic you can think of related to climate change, develops policy recommendations, and works with businesses on solutions to climate change. Their Web site offers a wealth of information, including a Global Warming Basics section, which can give you a great refresher on the essentials.

If you're really interested in digging more deeply into global warming issues, explore the Pew Center's reports. They investigate just about every aspect of the topic:

- **Business involvement:** Reports cover effective governance strategies, the effectiveness of emissions trading, and the market consequences of climate change.
- **International issues:** You can read about mitigation measures in China, transportation solutions in the developing world, the thorny issue of global fairness, and much more.
- **Science and impacts:** This section of the site houses reports that track the causes of global warming and examine the effects it's had on the planet.
- **Technology solutions:** These reports cover everything from long-range energy solutions to fixes people can implement right now, such as using efficient appliances and cleaner transportation options.

International Institute on Sustainable Development

This site (`www.iisd.org/climate`) offers a look into major connections between climate change and development (in both developing countries and industrialized countries) on a global level. It's pretty serious stuff and sort of heavy reading, but it's also very interesting, useful, and reliable information.

Scroll down the main page until you reach the Contents section (near the bottom). This section includes the following links (and more):

- **After Kyoto:** This page provides a glance into progress on Kyoto Protocol talks. *After Kyoto* refers to the next phase because Kyoto technically ends in 2012. This page provides breaking news, reports, and links to find out more about international work and foreign policy climate issues.
- **Vulnerability and adaptation:** This page gives an overview of the dangers that climate change puts countries in and what people can do in response. You can find a wealth of projects in and reports on countries around the world that are working to adapt to climate change.
- **Energy:** Development revolves around energy. This page gives a background on exactly how energy fits into sustainable development challenges and what it means in the context of climate change.

Index

• A •

• B •

• C •

• F •

• G •

• H •

• I •

• J •

• K •

• L •

• M •

• N •

• O •

• P •

• Q •

• R •

• S •

• T •

• U •

• Y •

• Z •